AF807192

Natural Gas Hydrate Management in Deepwater Gas Well

Zhiyuan Wang · Baojiang Sun ·
Yonghai Gao

Natural Gas Hydrate Management in Deepwater Gas Well

Science Press
Beijing

Springer

Zhiyuan Wang
School of Petroleum Engineering
China University of Petroleum
Qingdao, China

Baojiang Sun
School of Petroleum Engineering
China University of Petroleum
Qingdao, China

Yonghai Gao
School of Petroleum Engineering
China University of Petroleum
Qingdao, China

ISBN 978-981-15-6417-8 ISBN 978-981-15-6418-5 (eBook)
https://doi.org/10.1007/978-981-15-6418-5

Jointly published with Science Press
The print edition is not for sale in China (Mainland). Customers from China (Mainland) please order the print book from: Science Press.
ISBN of the Co-Publisher's edition: 978-7-03-064419-0

© Science Press and Springer Nature Singapore Pte Ltd. 2020
This work is subject to copyright. All rights are reserved by the Publishers, whether the whole or part of the material is concerned, specifically the rights of translation, reprinting, reuse of illustrations, recitation, broadcasting, reproduction on microfilms or in any other physical way, and transmission or information storage and retrieval, electronic adaptation, computer software, or by similar or dissimilar methodology now known or hereafter developed.
The use of general descriptive names, registered names, trademarks, service marks, etc. in this publication does not imply, even in the absence of a specific statement, that such names are exempt from the relevant protective laws and regulations and therefore free for general use.
The publishers, the authors, and the editors are safe to assume that the advice and information in this book are believed to be true and accurate at the date of publication. Neither the publishers nor the authors or the editors give a warranty, express or implied, with respect to the material contained herein or for any errors or omissions that may have been made. The publishers remain neutral with regard to jurisdictional claims in published maps and institutional affiliations.

This Springer imprint is published by the registered company Springer Nature Singapore Pte Ltd.
The registered company address is: 152 Beach Road, #21-01/04 Gateway East, Singapore 189721, Singapore

Foreword

Natural gas hydrate is one type of ice-like crystals converted from some gas with low molecular mass and volatile liquid under high-pressure and low-temperature. Natural gas hydrate, as a flow barrier, widely exists in the stages of deepwater oil and gas drilling, well test, production, gathering, and transportation, which seriously affects the operation safety and efficient development of deepwater oil and gas. The theory of natural gas hydrate flow assurance is an emerging discipline developing from the mid and late twentieth century. It is a new field for study on the formation, deposition, aggregation, and blockage of hydrate based on the subjects such as chemistry, fluid mechanics, solid mechanics, oil and gas engineering, chemical thermodynamics, heat and mass transfer, which provides scientific guidance and engineering foundation for the safe, economic and efficient development of deepwater oil and gas.

In this book, the theories and technologies for hydrate management during drilling and production in the deepwater gas well are systematically described in simple terms, so that scientific and technical workers at various levels can easily read and refer to the contents. This book analyzes the hydrate risks during drilling and production, introduces the hydrate phase equilibrium conditions, hydrate formation and decomposition of kinetics models; sets out the hydrate formation rate models with and without free water, reveals the formation mechanism of hydrate generated from condensed liquid film and droplets condensed in gas core; establishes the temperature and pressure fields model in deepwater gas wellbore, sets out the quantitative prediction method for hydrate formation area under various working conditions; experimentally analyzes the influence of hydrate volume fraction in wellbore on the drilling fluid rheology, establishes the multi-phase flow model in multicomponent system and reveals the influence of hydrate phase transition on multi-phase flow; based on the well multi-phase flow theory and interaction mechanism between hydrate particles, establishes the respective hydrate layer growth model in gas-liquid-solid three-phase flow, gas-solid two-phase flow and water-saturated single-phase flow of deepwater gas well; analyzes the existing hydrate management technologies, focuses on expounding the screening of inhibitors, parameters design and related calculation methods during inhibitor injection, then puts forward the hydrate blockage management method and risk early-warning method based on HBFW; based on the above study, develops the hydrate management software in deepwater gas well to achieve its popularization and application on site.

In this book, by combining the author's latest research findings, the systematic descriptions of theories and technologies for hydrate management not only inherit the achievements of the predecessors, but deepen and develop the hydrates flow assurance mechanisms and its management methods. Beyond all doubt, leading-edge and innovative opinions and understanding are involved. For example, it systematically expounds the mechanisms of hydrate formation, migration, deposition and blockage in the multi-phase flow of gas-dominated systems, builds the well multi-phase flow model of the multicomponent system considering the hydrate phase transition, reveals the influence of hydrate phase transition on multi-phase flow and puts forward the creative hydrate blockage management method based on HBFW, which greatly improves the hydrate prevention effect in deepwater well.

This book comprises novel contents and reasonable arrangement of chapters and sections, integrates the profound theories and industrial technology practice to facilitate the deep understanding of theories and technologies for hydrate management in deepwater well and provides guidance for the operation design and related construction of hydrate flow assurance. Consequently, it is an important work combing theory and application value in the field of deepwater oil and gas engineering.

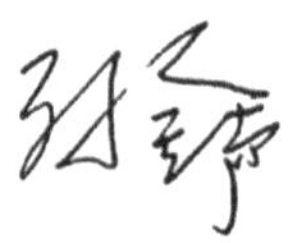

December 2019

Jinsheng Sun
School of Petroleum Engineering
China University of Petroleum
Qingdao, China

Preface

Energy supply is of great significance to the national economic and social development. Currently, the average consumption source of global energy is still dominated by oil, natural gas, and coal. The oil and gas resources on land in China keep entering the middle and later periods of development, and the external dependence of crude oil is increasing. The external dependence of crude oil in 2018 has reached 70.9%, which is much higher than the international oil safety threshold. Therefore, it is a great challenge for the oil and gas supplement.

The South China Sea is rich in oil and gas with geological reserves of 23–30 billion tons, 70% of which are located in deep water. Also, the national "13th Five-Year Plan" clearly proposed to strengthen the exploration and development of deepwater oil and gas, indicting deepwater oil and gas has become an important strategic area substituting the onshore oil-gas exploration and development. That is to say, the trend of oil and gas development "from onshore to offshore, from shallow sea to deep sea" is inevitable. However, the natural gas hydrates are easily generated near the mudline, then deposited and attached to the pipe wall, finally, the hydrate blockage occurs during deepwater drilling and gas production, which will cause the flow assurance and security problems, prolong the operation time, even lead to serious economic losses and oil-gas well abandonment. In brief, the hydrate flow barrier has become an important factor affecting the exploration safety of deepwater oil and gas, and has attracted extensive attention from scholars and operators at home and abroad.

Natural gas hydrate prevention and management in deepwater gas well involve the following contents: (1) Whether the hydrate is generated in wellbore—prediction on the hydrate formation area; (2) How the hydrate deposition and blockage are caused—evolution on hydrate deposition and blockage; (3) How to prevent and manage the hydrate blockage—hydrate blockage management method. The whole book is divided into seven chapters. Chapter 1—Overview on Hydrate Risks in Deepwater Oil and Gas Development—mainly introduces the risks of hydrate formation under varied working conditions such as drilling, well control, well testing, gas production, gathering and transportation, etc.; Chap. 2—Formation and Decomposition of Natural Gas Hydrate—mainly analyzes the hydrate formation and decomposition from two perspectives of thermodynamics and kinetics, provides the hydrate phase equilibrium model and calculation method of hydrate formation and decomposition rate; Chap. 3—Prediction for Natural Gas Hydrate Formation Area in Deepwater Gas Well—mainly introduces the temperature and pressure fields model of wellbore/pipeline in annular-mist flow, and puts forward the quantitative prediction method for hydrate formation area; Chap. 4—Influence of Hydrate Phase Transition on Multi-phase Flow in Deepwater Gas Well—mainly introduces the influence of hydrate formation on drilling fluid rheology, establishes the well annulus multi-phase flow model during drilling considering the hydrate phase transition and then analyzes the influence of hydrate phase transition on multi-phase flow in wellbore; Chap. 5—Mechanism and Prediction for Hydrate Deposition and Blockage in Deepwater Gas Well—mainly introduces the migration, gathering, deposition and blockage mechanisms of hydrate particles in multi-phase flow and establishes the hydrate blockage model; Chap. 6—Technologies for Hydrate Management in Deepwater Gas Well—mainly introduces the type of hydrate inhibitors and its screening method, carries out the optimization design of injection parameters and comes up with one new hydrate blockage prevention method based on HBFW; Chap. 7—Hydrate Management Software in Deepwater Gas Well and Case Analysis—mainly introduces the constituent modules of hydrate management software and its function, then analyzes the hydrate management scheme in deepwater gas well by case.

Publication of this book is supported by funds such as the major project of National Natural Science Foundation of China (51991360), the National Key R&D Program of China (973 Program, 2015CB251200), National Excellent Youth Fund (51622405), the Shandong Science Fund for Distinguished Young Scholars (JQ201716), “the Changjiang Scholars Program” of Education Ministry (Q2016135), Taishan Scholars and Specially-invited Expert of Shandong Province (ts201712018) and Talents Program.

This book is drafted and written by Zhiyuan Wang, Baojiang Sun and Yonghai Gao. Teachers and Ph.D. graduate students such as Chenwei Liu, Yang Zhao, Jianbo Zhang, Yangyang Zhang, Weiqi Fu, Jing Yu, Xiaohui Sun, Kai Du, Zheng Liu, Youqiang Liao, Chenru Zhou, Shikun Tong, Zhenggang Gong participate in the data collection, text correction and charts arrangement for this book. We express our heartfelt gratitude for their work on the occasion of the book’s publication.

The global deepwater oil-gas exploration and development are unusually active, so relevant theories and technologies for hydrate management, as the current research frontier, are constantly updated and improved. Due to the limited time, some omissions and mistakes are inevitable. Please do not hesitate to provide suggestions.

Qingdao, China

Zhiyuan Wang
Baojiang Sun
Yonghai Gao

Contents

Chapter 1
Overview on Hydrate Risks in Deepwater Oil and Gas Development

1.1 Engineering Background of Hydrate Risks

The South China Sea is rich in oil and gas with geological reserves of 23–30 billion tons, 70% of which are located in deepwater areas. Besides, the national "13th Five-Year Plan" clearly proposed to strengthen the exploration and development of deep-water oil and gas, indicting the deepwater oil and gas has become an important strategic replacement area of energies in China. During deepwater drilling, hydrate is likely to form in the low-temperature and high-pressure areas of wellbore or pipeline. The hydrate formation area during deepwater drilling and well completion is shown in Fig. 1.1, where the unceasing formation, gathering, and deposition of hydrate will cause hydrate blockage. Currently, in the deepwater gas well, the accidents of hydrate blockage have been detected in the Gulf of Mexico, British Atlantic, Brazilian Sea, and the South China Sea, etc., seriously affecting the operation schedule and causing gigantic economic losses. Hence, hydrate flow assurance has become an international key problem and more attention needs to be paid to the deepwater oil and gas development [1–4].

So far, most of the studies on hydrate blockage mainly focus on the production and transportation of oil and gas [5–12]. However, the researches of hydrate blockage in the deepwater gas well are still relatively few. Little knowledge or understanding of the evolution mechanism of hydrate flow barriers in the deepwater wellbore or pipeline poses great challenges to hydrate prevention and management on-site: the demand for hydrate management is uncertain [13, 14]. Moreover, the abuse of hydrate inhibitors causes high costs and heavy environmental pollution [15, 16]. Thus, delving into the evolution mechanism of hydrate flow barriers and developing efficient hydrate blockage management technologies are of great significance for the safety and efficiency of deepwater oil and gas exploitation. Purposefully, hydrate blockage risks under different working conditions of drilling, well control, testing, production, gathering and transportation shall be analyzed respectively.

© Science Press and Springer Nature Singapore Pte Ltd. 2020

Z. Wang et al., *Natural Gas Hydrate Management in Deepwater Gas Well*,
https://doi.org/10.1007/978-981-15-6418-5_1

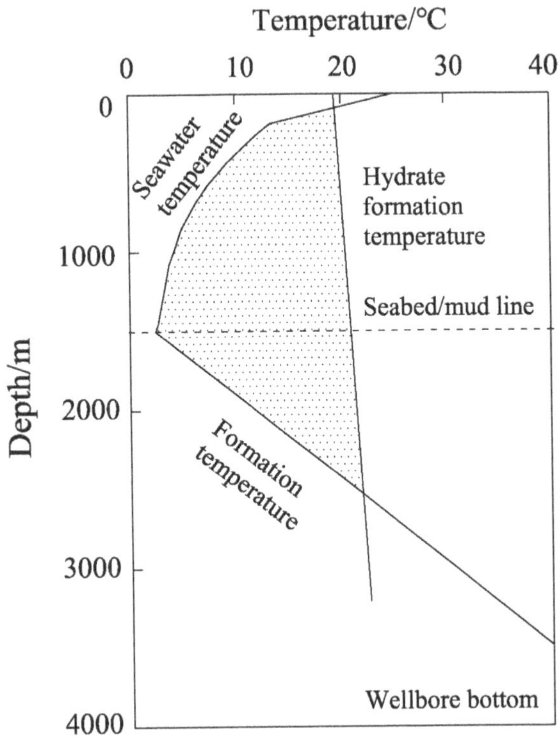

Fig. 1.1 Depth-temperature area of hydrate formation during deepwater drilling and well completion

1.2 Hydrate Risks During Deepwater Drilling and Well Control

Under deepwater high-pressure and low-temperature environment, the free water tends to contact the natural gas and form the hydrate. Further, during deepwater drilling and well control, natural gas enters the wellbore and makes its flow pattern transfer from the single-phase flow to the gas-liquid two-phase flow, finally become the gas-liquid-solid three-phase flow due to hydrate formation. Thus, the gas volume fraction in wellbore and increment of drilling fluid in mud pit will be reduced and the overflow will be delayed, making the early gas cutting be "concealed". Afterwards, hydrate decomposition during upward migration will greatly increase the gas volume fraction and increment of drilling fluid in the mud pit, making the later gas cutting be "abrupt" [17]. Both "concealness" and "abruptness" of gas cutting pose great challenges to the deepwater well control. Also, the continuous accumulation of solid hydrates will reduce the effective inner diameter of wellbore/pipeline, even lead to the hydrate blockage at pipe column, throttling line and the blowout preventer, which will bring great hazards to the deepwater well control [13, 18–22].

(1) Serious problems about operation and safety caused by hydrate deposition and blockage

The hydrate formation and deposition could cause the failure of well control and affect the construction progress. If the killing well or throttling pipeline is blocked by hydrate, the operation cycle cannot be performed; if the blowout preventer is blocked, it will fail in well closing and pressure monitoring of wellbore. Hydrate blockage could also increase the pipeline pressure and damage the pipelines and instruments, then the large pressure difference at positon of hydrate blockage may separate hydrate slug from the pipewall and grow the "moving shell" at one high speed, finally damage the pipeline or platform and even cause the accident of blowouts, explosions, etc. [23, 24]. Lately, the global hydrate blockage cases in the process of deepwater well control are shown in Table 1.1 [15, 16], which shows that the hydrate blockage has been regarded as one significant factor impacting on the safety of deepwater well control.

(2) High handling difficulty and costs after hydrate deposition and blockage

No doubt, it is difficult to remove the hydrate of wellbore/pipeline once hydrate blockage forms due to the complicated seabed and other unforeseen factors (like a typhoon, ocean current). On the one hand, there is still no mature and accurate method to determine the blockage position. On the other hand, the common methods

Table 1.1 Hydrate blockage cases in the process of deepwater well control

Case	Description of hydrate blockage accident	Accident hazards
1	The kick of ESS-107 well in the southeastern sea area of Brazil occurred during drilling in 2001, and the hydrate blocked the killing and throttling pipeline during well control	The operation progress was delayed by near 70 days
2	The kick of a well in the sea area of Gulf of Mexico occurred during drilling in 2000, and the work cycle cannot be performed, causing that killing pipelines, throttling pipelines, and blowout preventers were blocked	The operation progress was delayed by 6 days
3	The kick of a well (depth: 945 m) in the sea area of Gulf of Mexico occurred during drilling in 1989, and killing pipelines, throttling pipelines, and blowout preventers were blocked	The operation progress was delayed by 6 days
4	The kick of a well (depth: 350 m) at the west coast of U.S. occurred during drilling in 1986, and killing pipelines, throttling pipelines were blocked during well control, causing the repeated failure of well killing	The operation progress was delayed by 7 days

Table 1.2 Hydrate blockage cases during deepwater gas well test

Year	Region	Water depth/m	Hydrate formation depth/m	Hydrate blockage position/m	Time for blockage removal/h
2010	Gulf of Mexico, US	1722.89	0–2144	850–2144	>264
2012	Campos Basin, Brazil	2788	0–3000	1715–1798	4
2013–2015	South China Sea	1500	100–1920	Unknown	Unknown

for hydrate blockage removal are pressure reduction, heat injection and chemical injection. The drawbacks of these traditional methods—longer operation time for blockage removal, low efficiency and high costs—bitterly slow the operation and production progress and damage the economic benefits in deepwater oil and gas development [23, 25].

1.3 Hydrate Risks During Deepwater Gas Well Test

The hydrate blockage cases during the deepwater gas well test are shown in Table 1.2 [26–28], which indicates hydrate formation and blockage could also endanger the well test. The main causes of hydrate formation during deepwater gas well test include:

(1) Fast heat exchange between internal and external of pipe column during well test, and high reduction rate of fluid temperature;
(2) Increasing pressure of pipe column after well closing;
(3) Higher pressure and lower temperature of pipe column during well opening;
(4) The sharp drop of temperature caused by throttling effects.

Besides, the factors that accelerate the hydrate formation include:

(1) The higher flow rate of the air;
(2) Large pressure fluctuation;
(3) Adequate gas-water contact;
(4) Existence of hydrate "seed crystal";
(5) Existence of acid gas such as H_2S.

1.4 Hydrate Risks During Deepwater Oil and Gas Production

The hydrate blockage during deepwater oil and gas exploration will also damage the equipment and threaten personal safety. Table 1.3 shows hydrate blockage during

Table 1.3 Accidents of hydrate blockage

Case	Description of hydrate blockage accident	Accident hazards
1	Three hydrate blockage accidents in the seabed pipeline of North Sea during 1990–1993	Three people died and the economic losses reached more than US$ 700 million
2	The hydrate blockage of wellbore and pipeline in Qatar oil and gas field happened repeatedly during 2011–2013	The well was closed and production was halted for months. The economic losses reached up to US$ 10 million per day
3	The pipeline with the length of 11.5 km in Tommeliten Gamma oil and gas field of the North Sea was forced to be closed due to hydrate blockage	seventeen hydrate blockages were found during blockage removal

oil and gas production could extend the operation cycle, increase the operating costs and even cause the involved accidents [18, 29, 30].

In the process of fluid migration from downhole to the wellhead, both temperature and pressure along the pipeline gradually reduce (the reduction rate of temperature is related to the gas productivity of well [31]). Consequently, the wellbore fluid would be under the condition of high-pressure and low-temperature, which makes hydrate formation easier. Besides, the temperature and pressure would decrease sharply due to the Joule-Thomson effect when the fluid flows through the wellhead and throttling valve, and hydrate risks grow when the fluid temperature drops to the hydrate phase equilibrium condition.

Under the conditions of well closing and halt production, the well pressure gradually grows to the closing pressure. The free water drops to the downhole under gravity and the natural gas in wellbore would become the status of water saturations. Meanwhile, the well temperature decreases continuously under the action of an external low-temperature environment. Once the temperature is lower than the water dew point, part free water would be condensed from the saturated natural gas. Thus, under the combined action of low-temperature seawater and well closing pressure, hydrate formation risk in a wellbore may occur and gradually increase with the well closing time. Notice that when the deepwater gas well is reopened, the high-pressure and low-temperature in well, well disturbance are advantageous to aggravate the hydrate formation risk.

1.5 Management Measures for Hydrate Risks

At present, the hydrate risks management measures in the deepwater pipelines mainly include dehydration, gas components separation and inhibitors injection, etc. Dehydration refers to removing and separating the free water produced from the reservoir before it enters into the transportation pipeline, which needs to mount the related equipment and manifolds and is mostly used in the conditions of large liquid yield.

However, this operation can only remove the free water and several water vapor still exist after dehydration. Thus, the pipeline is a water-saturated gas system, and the condensation of free vapor (namely, condensed water) could still be observed. Light hydrocarbon components could be separated from the mixed gas by depressurization separation, which requires continuous compression and pumping so that the operation costs and operation difficulty are increasing. Inhibitor injection would move the hydrate phase equilibrium conditions toward the direction of high-pressure or low-temperature to prevent the hydrate formation.

For the hydrate flow barriers in wellbore/pipeline, there are many methods to cope with hydrate blockage on-site such as local heating, depressurization, mechanical clean and inhibitors injection, etc. The local heating method would obtain one higher pressure in the pipeline due to hydrate decomposition. The depressurization method must be applied above the water freezing point to avoid the ice blockage. The mechanical clean method needs the additional clean equipment and has slow removal efficiency. Therefore, excessive injection of inhibitors is now mostly used for the full hydrate inhibition. However, the method has high requirements for the storage space and injection equipment, and has the disadvantages of high costs and toxicity, which further increase the operational risks.

References

1. Sun J, Ning F, Lei H et al (2018) Wellbore stability analysis during drilling through marine gas hydrate-bearing sediments in Shenhu area: a case study. J Petrol Sci Eng 170:345–367
2. Ning F (2005) Study on stratum well wall stability of natural gas hydrate. China University of Petroleum
3. Wang R, Ning F, Liu T et al (2017) Dynamic simulation on free methane gas forming hydrates in wellbore. Acta Petrol Sinica 38(8):963–972
4. Ruppel C, Boswell R, Jones E (2008) Scientific results from Gulf of Mexico Gas Hydrates Joint Industry Project Leg 1 drilling: introduction and overview. Mar Petrol Geol 25(9):819–829
5. Wang Z (2018) Hydrates blockage detection system in gas conveying pipeline with pressure wave propagation method. Dalian University of Technology
6. Gong Q, Ma G, Pan Z et al (2017) Study on calculation of settlement rules of natural gas hydrate in pipe conveying process. J Liaoning Univ Petrol Chem Technol 37(3)
7. Yang J, Pu C (2004) Study on formation theory model and blockage prediction of natural gas hydrates in gas gathering pipeline. Nat Gas Geosci 15(6):660–663
8. Chen K, Liu J, Zhang L et al (2005) Study on hydrate blockage mechanism and working conditions prediction during natural gas conveying pipe cleaning. Nat Gas Ind 25(4):134–136
9. Pu C, Song X, Yang J (2005) Study on natural gas hydrate formation and blockage prediction for gas gathering pipeline. Technol Superv Petrol Ind J Agency 21(5):71–74
10. Liu Enbin, Li C, Peng S et al (2006) Pipeline blockage detection technology based on pressure wave propagation method. Nat Gas Ind 26(4):112–114
11. Li P, Fu G, Li A et al (2012) Engineering practice on solving hydrates blockage in pipeline with pipe cleaning method: petrochemical industry innovation green sustainable development–Collected papers of petrochemical industry thematic forum of the eighth Ningxia young scientists forum
12. Ruan C, Shi B, Ding L et al (2016) Study progress of growth and pipe blockage rules of natural gas hydrate. Oil Gas Storage Transp 35(10):1027–1037

13. Wang Z (2009) Study on annular multiphase flow pattern transfer mechanism containing natural gas hydrate phase transition. China University of Petroleum (East China)
14. Ye J, Liu Z, Luo J et al (2015) Technical difficulties and countermeasures for deepwater drilling well control in the South China sea. Oil Drill Prod Technol 1:139–142
15. Barker WJ, Gomez KR (1989) Formation of hydrates during deepwater drilling operation. J Petrol Technol 41:297–301
16. Zhao Y (2017) Study on deposition and blockage of natural gas hydrates in deepwater wellbore multiphase flow system. China University of Petroleum (East China)
17. Wang Z, Sun B (2014) Deepwater gas kick simulation with consideration of the gas hydrate phase transition. J Hydrodyn 26(1):94–103
18. Lysne D, Larsen R, Thomsen ÅK et al (1995) Hydrate problems in pipelines: a study from norwegian continental waters. The fifth international offshore and polar engineering conference. International Society of Offshore and Polar Engineers
19. Dendy ES (2003) Fundamental principles and applications of natural gas hydrates. Nature 426(6964):353–363
20. Taylor CJ, Miller KT, Koh CA et al (2007) Macroscopic investigation of hydrate film growth at the hydrocarbon/water interface. Chem Eng Sci 62(23):6524–6533
21. Licence P (2007) Clathrate hydrates of natural gases. In: Dendy Sloan E, Koh Carolyn A. Chemical industries series, 3rd edn. CRC Press 87(13):3158
22. Van Der Waals JH, Platteeuw JC (2007) Clathrate solutions. Advances in Chemical Physics
23. Jassim E, Abdi MA, Muzychka Y (2010) A new approach to investigate hydrate deposition in gas-dominated flowlines. J Nat Gas Sci Eng 2(4):163–177
24. Gao Y, Chen Y, Zhao X et al (2018) Risk analysis on the blowout in deepwater drilling when encountering hydrate-bearing reservoir. Ocean Eng 170:1–5
25. Sloan ED (2005) A changing hydrate paradigm—from apprehension to avoidance to risk management. Fluid Phase Equilib 228–229(3):67–74
26. Reyna EM, Stewart et al (2001) Case history of the removal of a hydrate plug formed during deep water well testing. In: SPE/IADC drilling conference. Society of Petroleum Engineers
27. Chen SM, Gong WX, Antle G (2008) DST design for deepwater wells with potential gas hydrate problems. In: Proceedings of offshore technology conference. Society of Petroleum Engineers Offshore Technology Conference
28. Yang Z, Wang Z, Jing Y et al (2017) Hydrate plug remediation in deepwater well testing: a quick method to assess the plugging position and severity. In: SPE annual technical conference and exhibition. Society of Petroleum Engineers SPE Annual Technical Conference and Exhibition
29. Austvik T, Xiaoyun LI, Gjertsen LH (2010) Hydrate plug properties: formation and removal of plugs. Ann N Y Acad Sci 912(1):294–303
30. Mohamed NA (2014) Avoiding gas hydrate problems in Qatar oil and gas industry: environmentally friendly solvents for gas hydrate inhibition. World Academy of Science, Engineering and Technology International Journal 8
31. Jin S, Zhang C, Meng W et al (2015) Formation risks and prevention measures of natural gas hydrates in drilled well 17-2 of deepwater gas field in Lingshui. China Offshore Oil and Gas 27(4):93-101

Chapter 2
Formation and Decomposition of Natural Gas Hydrate

The mechanism and mathematical model on the formation and decomposition of Natural Gas Hydrate (NGH) are the basis for studying the theories and technologies for hydrate prevention and management. This chapter briefly introduces the structure and formation of NGH, then analyzes the mechanism of hydrate formation and decomposition in thermodynamics and kinetics, finally provides the calculation methods of hydrate formation and decomposition rate.

2.1 Structure and Formation of NGH

Under the high-pressure and low-temperature, one ice-like crystal known as "clathrate hydrate" or simply as "hydrate" could form by the contact between partial gas with low molecular mass or volatile liquid (guest molecule) and free water (host molecule). Thus far, more than one hundred substances, including methane, ethane, ethylene, propane and carbon dioxide, are expected to form the desired hydrate. Apparently, as the mixture of these components, the natural gas would form NGH [1–3]. For the hydrate structure, there are the cage lattice (named "water cage", including small and large space) constructed by the hydrogen bond of host molecules, and guest molecules are optionally filled in the water cage. The interaction force between host and guest molecules is Van der Waals's force.

Under the specific temperature and pressure, the water cage could constitute the pentagonal dodecahedron, the tetrakaidecahedron composed of 12 pentagons and 2 hexagons, or the polyhedron composed of 12 pentagons and 4 hexagons. Afterwards, the stable hydrate crystals are formed by matching the water cage and guest molecules. The cage structures of hydrate are shown in Fig. 2.1 [1].

As the host and guest molecules are more stable by Van der Waals, hydrate crystals continue to grow, absorb and attract mutually to form the crystal particles of hydrate [4]. With the continuous gathering of crystal particles, the crystal block volume of hydrate becomes larger and finally forms the solid-gas hydrate. The process of hydrate formation, including four stages as follows, is shown in Fig. 2.2 [5].

© Science Press and Springer Nature Singapore Pte Ltd. 2020

Z. Wang et al., *Natural Gas Hydrate Management in Deepwater Gas Well*,
https://doi.org/10.1007/978-981-15-6418-5_2

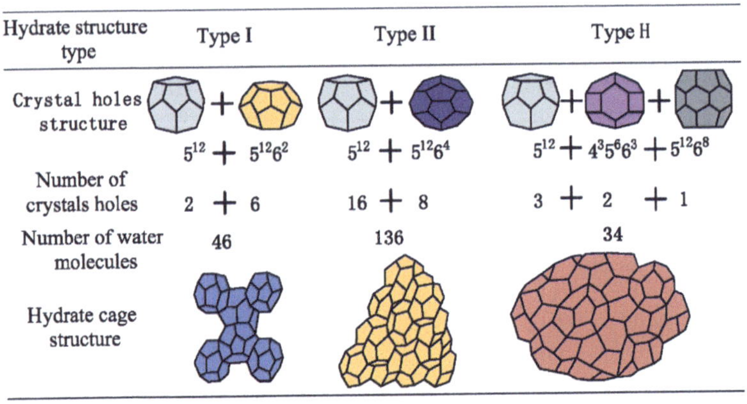

Fig. 2.1 Cage structures of hydrate

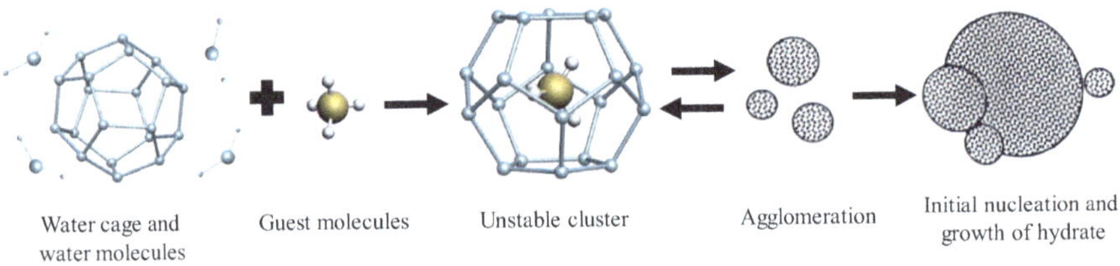

Fig. 2.2 Hydrate formation process

(1) Initial conditions: both temperature and pressure are located in the hydrate formation area, in which there are no dissolved gas molecules into the free water.
(2) Unstable cluster: gas molecules start to dissolve into the free water to form an unstable cluster immediately.
(3) Agglomeration: unstable clusters agglomerate together by face contact and have high irregularity.
(4) Initial nucleation and growth of hydrate: hydrate crystals start to grow as soon as the agglomeration exceeds a certain threshold value.

Currently, three crystal structures of NGH: type I, type II and type H have been confirmed, of which structural parameters are as shown in Table 2.1 [6].

The stability of hydrate crystal structure is determined by type and size of guest molecules, and the stable hydrate structure would be obtained only if the space of water cage match the size of guest molecules (ordinarily the diameter ratio of guest molecules and water cage is near 0.9). However, the gases with high solubility such as ammonia, hydrogen chloride cannot form the hydrate regardless of its molecule size. Common guest molecules mainly include hydrocarbon components (CH_4, C_2H_6, C_3H_8, *i*-C_4H_{10}, *n*-C_4H_{10}, etc.), acidic component CO_2, H_2S and inert components (N_2, Ar, Kr, Xe), etc.

Table 2.1 Structural parameters of three hydrate crystal structures

Structural types of hydrate		Type I	Type II	Type H
Crystal structure of hydrate		Body centered cube	Diamond cube	Simple hexahedron
Small water cage	Structure	5^{12}	5^{12}	5^{12}, $4^3 5^6 6^3$
	Diameter (nm)	7.82	7.8	7.6
Large water cage	Structure	$5^{12}6^2$	$5^{12}6^4$	$5^{12}6^8$
	Diameter (nm)	8.66	9.36	10.4
Number of small water cages per hydrate crystal cell		2	16	3, 2
Number of large water cages per hydrate crystal cell		6	8	1
Number of water molecules per hydrate crystal cell		46	136	34
Ratio of small water cages and water molecules per hydrate crystal cell		1/23	2/17	3/34, 1/17
Ratio of large water cages and water molecules per hydrate crystal cell		3/23	1/17	3/17
Molecular formula of hydrate crystal cell		$S_2L_6{\cdot}46H_2O$	$S_{16}L_8{\cdot}136H_2O$	$S_3S_2L_1{\cdot}34H_2O$

The structure types of hydrate are mainly determined by the size of guest molecules, which is also affected by factors such as the shape of guest molecules, temperature and pressure conditions, and the existence of a hydrate promoter. Smaller guest molecules could enter the various holes of hydrate crystals, and exert different stability effect. In practice, larger guest molecules only stabilize the holes of type II, and exert much stability than that of smaller guest molecules. Therefore, the propane, butane, and *i*-butane tend to form type II of the hydrate. Nevertheless, with the help of the hydrate formation experiments of mixed methane, ethane and propane, Holder and Angert [7] found that both various gas composition, temperature, and pressure would transfer the hydrate structural type, which makes it difficult to judge the hydrate structural type in the mixed gas system. For "hydration number", the host and guest molecules will maintain this fixed proportional relation when water cages are fully filled by guest molecules, which is 5.75 in type I and 5.67 in type II of structure. In fact, the guest molecules only fill partial water cages with a filling rate between 0.9 and 1.0.

The matching size between guest molecules and water cages is shown in Fig. 2.3, and the diameter proportion of guest molecules to water cages is shown in Table 2.2 [3, 8]. From Fig. 2.3 and Table 2.2, the following can be seen:

(1) Gas molecules (He, Ne, etc.) less than 3.5Å in diameter cannot stabilize the water cages and form hydrates; gas molecules larger than 7.5Å in diameter cannot fill into any water cages and form hydrates due to the limited size of water cage;

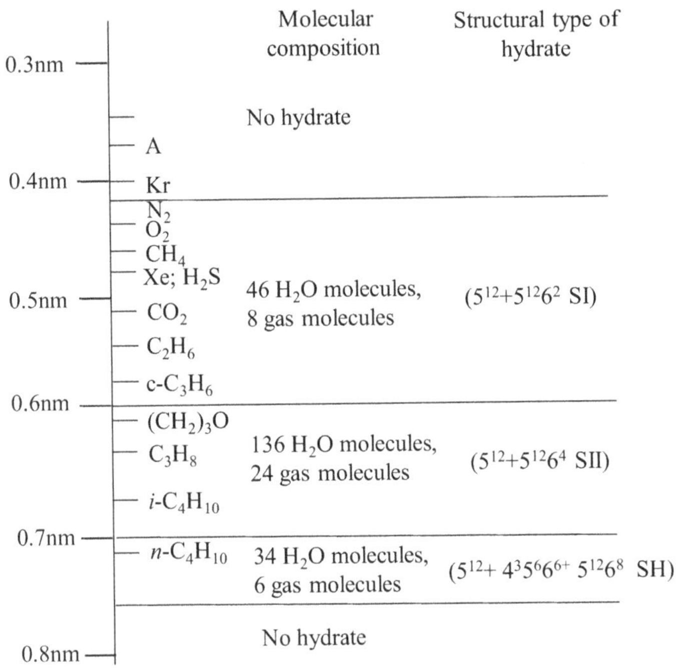

Fig. 2.3 Matching relation of size between guest molecules and water cages

Table 2.2 Diameter proportion of guest molecules to water cages

Guest molecules		Diameter proportion of guest molecule to water cage			
Name	Diameter (Å)	Type I		Type II	
		5^{12}	$5^{12}6^2$	5^{12}	$5^{12}6^4$
Ne	2.97	0.604	0.516	0.606	0.459
Ar	3.8	0.772	0.660	0.775#	0.599#
Kr	4.0	0.813	0.694	0.816#	0.619#
N_2	4.1	0.833	0.712	0.836#	0.634#
O_2	4.2	0.853	0.729	0.856#	0.649#
CH4	4.36	0.886	0.757	0.889	0.675
Xe	4.58	0.931	0.795	0.934	0.708
H_2S	4.58	0.931	0.795	0.934	0.708
CO_2	5.12	1.041	0.889	1.044	0.792
C_2H_6	5.5	1.118	0.955	1.122	0.851
C_3H_8	6.28	1.276	1.090	1.280	0.971#
i-C_4H_{10}	6.5	1.321	1.128	1.325	1.005#
n-C_4H_{10}	7.1	1.443	1.232	1.447	1.098

(2) Gas molecules like ethane alone fill into the large water cages $5^{12}6^2$ of type I, and propane and *i*-butane, etc. alone fill into the large water cages $5^{12}6^4$ of type II;
(3) The gas components such as methane, hydrogen sulfide and carbon dioxide could enter the small water cages 5^{12} and large water cages $5^{12}6^2$ of type I. That is why the structure type of involved hydrate is often named as $5^{12}+5^{12}6^2$;
(4) The smaller monatomic and diatomic gases such as argon, krypton, nitrogen and oxygen could fill into the small water cages 5^{12} and large water cages $5^{12}6^4$ of type II to form the hydrate structure of $5^{12}+5^{12}6^4$;
(5) The filling uncertainty of gas molecules and the filling fraction between 0.9 and 1.0 make hydrates non-stoichiometric.

As one type of mixed gas, Natural gas is composed of H_2S, CH_4, CO_2, C_2H_6, C_3H_8, *i*-C_4H_{10}, and *n*-C_4H_{10}, etc. These guest molecules could form the type I and type II of hydrate in theory, but only the more stable structure could keep its stability. Note that the largest gas component determines the structure type of hydrate.

2.2 Formation of NGH

2.2.1 *Hydrate Phase Equilibrium Condition*

Based on phase equilibrium theory, the thermodynamic conditions of gas hydrate phase equilibrium generally refer to the temperature and pressure in the gas-water-hydrate coexistent system, so the essence of gas-water-hydrate three-phase equilibrium is calculating the temperature and pressure for hydrate formation.

The prediction methods of temperature and pressure for hydrate formation include empirical graphical method, phase equilibrium constant method (Katz method) and statistical thermodynamic method. The statistical thermodynamic method could obtain accurate results and lay a foundation for the subsequent hydrate formation prediction, but the calculation process is complex. In 1959, Van der Waals and Platteeuw [4] derived the VdW-P model by combining the hydrate structural characteristics and gas isothermal adsorption theory proposed by Langmuir, and it is the theoretical basis for statistical thermodynamics. Afterwards, Parrish [9] gave the formula of Langmuir constant, and popularized the VdW-P model to the multi-component gas. Ng, Robinson [10] and Holder, etc. [7] revised the Langmuir constant to obtain higher prediction precision of hydrate phase equilibrium.

2.2.1.1 Empirical Graphical Method

Currently, the empirical graphical method is the simplest prediction method of hydrate phase equilibrium conditions. The phase equilibrium curves of hydrate formed by methane and natural gas with different relative densities are shown in Fig. 2.4, in which it presents the hydrate formation area above the curve, but the

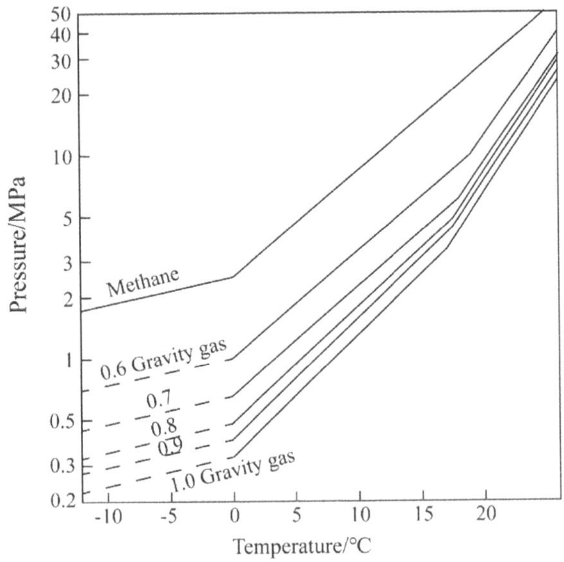

Fig. 2.4 Phase equilibrium curves of hydrate formed by methane and natural gas with different relative densities

unstable area of hydrate below the curve. It can be seen that higher pressure and lower temperature are conducive to the hydrate formation. For the sake of convenient computation, the hydrate phase equilibrium curve in the figure could be fitted to be polynomial shown in Table 2.3 [11].

The relation between reference pressure and gas pressure is shown in Eq. 2.1. If the relative density of natural gas is known, the hydrate formation temperature at one certain pressure could be found in Fig. 2.4. For the other relative density of natural gas, the temperature and pressure for hydrate formation could be determined by linear interpolation.

$$P = 10^{-3} \times 10^{P^*} \tag{2.1}$$

where P is the gas pressure, MPa; P^* stands for the reference pressure, MPa; T represents the gas temperature, °C.

Table 2.3 Fitted formula of hydrate phase equilibrium curve

Relative density	Fitted expression
0.6	$P^* = 3.009796 + 5.284026 \times 10^{-2}T - 2.252739 \times 10^{-4}T^2 + 1.1511213 \times 10^{-5}T^3$
0.7	$P^* = 2.814824 + 5.019608 \times 10^{-2}T - 3.722427 \times 10^{-4}T^2 + 3.781786 \times 10^{-6}T^3$
0.8	$P^* = 2.70442 + 5.82964 \times 10^{-2}T - 6.639789 \times 10^{-4}T^2 + 4.008056 \times 10^{-5}T^3$
0.9	$P^* = 2.613081 + 5.715702 \times 10^{-2}T - 1.871161 \times 10^{-4}T^2 + 1.93562 \times 10^{-5}T^3$
1	$P^* = 2.527849 + 0.0625T - 5.781363 \times 10^{-4}T^2 + 3.069745 \times 10^{-5}T^3$

2.2.1.2 Chen–Guo Model

Chen-Guo model [12] is one statistical thermodynamic model based on the dual-process kinetics of hydrate formation. The chemical reaction of hydrate formation is shown in Eq. 2.2. The quasi chemical equilibrium and physical adsorption equilibrium of gas molecules coexist when the hydrate is formed by single-component gas. For one thing, constrained by the chemical equilibrium, the chemical reaction could be described as Eq. 2.3. For another, the physical absorption equilibrium of gas molecules could reduce the chemical potential of the hydrate. The chemical potential of hydrate and gas could be shown in Eqs. 2.4 and 2.5 respectively.

$$H_2O + \lambda_2 A \longrightarrow A_{\lambda_2}H_2O \tag{2.2}$$

$$\mu_B = \mu_W + \lambda_2 \mu_G \tag{2.3}$$

$$\mu_B = \mu_B^0 + \lambda_1 RT \ln(1 - \theta) \tag{2.4}$$

$$\mu_G = \mu_G^0(T) + RT \ln f \tag{2.5}$$

where λ_2 is the hole constant of gas molecule A; μ_B stands for the chemical potential of hydrate, J/mol; μ_W represents the chemical potential of water, J/mol; μ_G represents the chemical potential of gas, J/mol; μ_B^0 refers to the chemical potential of pure hydrate, J/mol; λ_1 shows the number of connecting holes formed by water molecules; θ shows the fraction of connecting holes occupied by gas molecules, %; μ_G^0 refers to the chemical potential of gas in standard condition, J/mol.

By substituting the Eqs. 2.4 and 2.5 into Eq. 2.3, the following equation can be obtained:

$$\mu_B^0 + \lambda_1 RT(1 - \theta) = \mu_W + \lambda_2 \left[\mu_G^0(T) + RT \ln f\right] \tag{2.6}$$

Setting $f^0 = \exp\left[\frac{\mu_B^0 - \mu_W - \lambda_2 \mu_G^0(T)}{\lambda_2 RT}\right]$, Eq. 2.6 can be expressed as

$$f = f^0(1 - \theta)^a \tag{2.7}$$

where $a = \lambda_1/\lambda_2$; a takes 1/3 for type I of hydrate and 2 for type II of the hydrate.

When $\theta = 0$, the above equation would grow the Eq. 2.8. Therefore, f^0 indicates the gas phase fugacity in the state of pure hydrate phase equilibrium.

$$f = f^0 \tag{2.8}$$

According to the classical thermodynamics,

$$\mu_B^0 = A_B^0 + PV_B^0 \tag{2.9}$$

$$\mu_B^0 - \mu_W = \Delta A + P\Delta V - RT\ln a_W \tag{2.10}$$

so, $\mu_B^0 - \mu_W$ in Eq. 2.6 can be expressed as

$$\mu_B^0 - \mu_W = \Delta A + P\Delta V - RT\ln a_W \tag{2.11}$$

It is assumed that the free energy of Helmholtz is only related to the temperature, and the molar volume is a constant when the pressure is less than 100 MPa due to the small compressibility of water and solid hydrate. Hence, $\Delta A = A_B^0 - A_W$ could be deemed as the temperature function, but ΔA could be considered as a constant. So, the Eq. 2.8 can be expressed as

$$f^0 = f_T^0 f_P^0 f_{aW}^0 \tag{2.12}$$

$$f_P^0 = \exp\left(\frac{\beta P}{T}\right) \tag{2.13}$$

$$f_{aW}^0 = a_W^{\frac{-1}{\lambda_2}} \tag{2.14}$$

where $\beta = \frac{\Delta V}{\lambda_2 R}$ is only related to the structural type of hydrate, which takes 4.24 K/MPa for type I of hydrate and 10.224 K/MPa for type II of hydrate;a_W refers to water activity in the water-rich phase.

Associate f_T^0 into the temperature function in the form of Antoine equation:

$$f_T^0 = a\exp\left(\frac{b}{T-c}\right) \tag{2.15}$$

where *a*, *b* and *c* stand for the regression coefficients represented in Table 2.4.

Table 2.4 Regression coefficients of *a*, *b* and *c*

Gas	Type I of hydrate			Type II of hydrate		
	a /10^{-11}MPa	b/K	c/K	a /10^{-24}MPa	b/K	c/K
CH_4	1584.4	−6591.4	27.04	5.2602	−12955	4.08
C_2H_6	47.5	−5465.6	57.93	0.0399	−11491	30.4
C_3H_8	100	−5400	55.5	4.1023	−12312	30.2
i-C_4H_{10}	1.0	0	0	4.5138	−12850	37.0
n-C_4H_{10}	1.0	0	0	3.5907	−12312	39.0
H_2S	4434.2	−7540.6	31.88	3.2794	−13523	6.7
CO_2	963.72	−6444.5	36.67	3.4474	−12570	6.79

The hydrate formed by mixture gas could be regarded as a solid solution composed of several hydrates. Because the molar volumes of different gas hydrates with the same structure are very close, the mixture of several hydrates is approximately one normal solution. Ignoring the interaction between different gas molecules in hydrate, the thermodynamic model of hydrate could be

$$\begin{cases} f_i = x_i f_i^0 \left(1 - \sum_j \theta_j\right)^a \\ \sum_j \theta_j = \frac{\sum_j f_j C_j}{1+\sum_j f_j C_j} \\ \sum_i x_i = 1 \end{cases} \tag{2.16}$$

where f_i is fugacity of component i, MPa; f_i^0 is fugacity of component i in pure basic hydrate, MPa; θ_j refers to the fraction of connecting hole occupied by gas molecules j; x_i shows the mole proportion of hydrate formed by gas molecules i to the mixed hydrate.

2.2.1.3 VdW-P Model

By combining the hydrate structural characteristics, gas isothermal adsorption theory proposed by Langmuir, and the treating method of statistical thermodynamics, Van der Waals and Platteeuw [4] derived the VdW-P model. During hydrate formation, the thermodynamic model of gas-liquid-solid three-phase equilibrium usually describes the two parts of the hydrate phase and water-rich phase. Generally, considering the water as the reference component, and introducing the chemical potential μ_β of water in hydrate phase (β phase) as the reference status, the following equations could be concluded in case of hydrate phase equilibrium:

$$\Delta\mu_{\mathrm{H}} = \mu_\beta - \mu_{\mathrm{H}} = RT\sum_{i=1}^{2} v_i \ln(1 - \sum_{j=1}^{N_{\mathrm{C}}} \theta_{ij}) \tag{2.17}$$

$$\theta_{\mathrm{ij}} = \frac{C_{ij} f_j}{1 + \sum_{j=1}^{N_{\mathrm{C}}} C_{ij} f_j} \tag{2.18}$$

$$C_{ij} = \frac{A_{ij}}{T} \exp\left(\frac{B_{ij}}{T}\right) \tag{2.19}$$

where T shows the system temperature, °C; v_i stands for the number of i-type holes in the hydrate phase; θ_{ij} represents the proportion of guest molecules j to i-type holes, %, which could be solved by Eq. 2.18; C_{ij} represents the Langmuir constant of guest molecules j in i-type holes, which could be solved by Eq. 2.19; f_j refers to the fugacity of guest molecules j in each phase, MPa; N_{C} is the number of the molecules in the mixture for hydrates formation; A_{ij}, B_{ij} show the empirical constant.

When gas hydrate and non-gas hydrate have stayed in equilibrium:

$$\mu_{\mathrm{W}} + RT\sum_{i=1}^{2} \nu_i \ln\left(1 - \sum_{j=1}^{N_{\mathrm{C}}} \theta_{ij}\right) = \mu_{\mathrm{W}}^{0} + RT\ln(f_{\mathrm{W}}/f_{\mathrm{W}}^{0}) \tag{2.20}$$

where μ_{W} is the chemical potential of water in the water-rich phase, J/mol; f_{W} is the fugacity of water in water-rich phase, MPa; μ_{W}^{0} represents the chemical potential of pure water under T and P in reference states, J/mol; f_{W}^{0} represents the fugacity of pure water under T and P in reference states, MPa.

Besides, the chemical potential of water could be also expressed as

$$\frac{\mu_{\mathrm{W}} - \mu_{\mathrm{W}}^{0}}{RT} = \frac{\Delta\mu_0}{RT_0} - \int_{T_0}^{T} \frac{\Delta H_0 + \Delta C_{\mathrm{P}}(T - T_0)}{RT^2}\mathrm{d}T + \int_{P_0}^{P} \frac{\Delta V}{RT}\mathrm{d}P \tag{2.21}$$

According to the Eqs. 2.20 and 2.21, the hydrate phase equilibrium conditions can be obtained as below:

$$\begin{aligned} &\frac{\Delta\mu_0}{RT_0} - \int_{T_0}^{T} \frac{\Delta H_0 + \Delta C_{\mathrm{P}}(T - T_0)}{RT^2}\mathrm{d}T + \int_{P_0}^{P} \frac{\Delta V}{RT}\mathrm{d}P \\ &= \ln(f_{\mathrm{W}}/f_{\mathrm{W}}^{0}) - \sum_{i=1}^{2} \nu_i \ln\left(1 - \sum_{j=1}^{N_{\mathrm{C}}} \theta_{ij}\right) \end{aligned} \tag{2.22}$$

where $\Delta\mu_0$ is the chemical potential difference of water in hydrate structure and pure water under standard condition, J/mol; T_0, P_0 refers to the temperature (K) and pressure (MPa) under standard condition respectively, and $T_0 = 273.15$ K, $P_0 =$ 0 MPa; ΔH_0, ΔV, ΔC_{P} represent the specific enthalpy difference (J/mol), specific volume difference (m^3/kg) and specific heat difference (J/(kg·K)) between hydrate structure and pure water respectively; $\ln(f_{\mathrm{W}}/f_{\mathrm{W}}^{0}) = \ln x_{\mathrm{W}}$, if the hydrate inhibitor is added, there is $\ln(f_{\mathrm{W}}/f_{\mathrm{W}}^{0}) = \ln(y_{\mathrm{W}} x_{\mathrm{W}})$; x_{W}, y_{W} show the mole fraction and activity coefficient of water in the water-rich phase.

If the system pressure is completely known, the solution procedure of phase equilibrium temperature of hydrate are as follows:

(1) Input the basic data and phase equilibrium temperature of hydrate under one pressure;
(2) Calculate the fugacity of gas-phase with PR Equation;
(3) Determine the structural type of hydrate and calculate the constants of θ_{ij} and C_{ij};
(4) Solve the phase equilibrium temperature of hydrate adopting the iterative approach.

Likewise, if the system temperature is fully known, the solution method for hydrate phase equilibrium pressure is the same as the steps above. After the hydrate formation temperature-pressure are determined, it could be judged whether to form the solid-phase hydrate or not under the conditions of one certain temperature, pressure and gas-liquid phase.

2.2.2 *Hydrate Formation Dynamics*

The evolution process of flow barriers in wellbore/pipeline is impacted by hydrate formation kinetics, and it used to be that the hydrate formation rate is mainly controlled by three mechanisms of intrinsic kinetics, heat transfer, and mass transfer. Based on such mechanisms, corresponding models of hydrate formation rates are established.

2.2.2.1 Intrinsic Kinetic Model

The intrinsic kinetic model of hydrate holds that the main factor determining the hydrates growth is the guest molecules and factors such as gas components, water salinity, and inhibitor components are the intrinsic kinetic conditions during hydrate growth. Vysniauskas and Bishnoi model [13], as an well-accepted intrinsic kinetic model, holds that the driving force of hydrate formation mainly originates from the fugacity difference of gas, and the hydrate formation is subject to following steps:

(1) Initial cluster reaction:

$$\mathrm{M} + \mathrm{H_2O} + (\mathrm{H_2O})_{y-1} \rightleftharpoons \mathrm{M} \cdot (\mathrm{H_2O})_y \tag{2.23}$$

(2) Formation of the hydrate crystal nucleus with critical dimensions:

$$\mathrm{M} + \mathrm{H_2O} + \mathrm{M}(\mathrm{H_2O})_x \rightleftharpoons \mathrm{M} \cdot (\mathrm{H_2O})_c \tag{2.24}$$

(3) Growth of hydrate crystal:

$$\mathrm{M} + \mathrm{H_2O} + \mathrm{M}(\mathrm{H_2O})_m \rightleftharpoons \mathrm{M} \cdot (\mathrm{H_2O})_n \tag{2.25}$$

The induction time of nucleation depends on the generation probability of cluster with critical dimensions, which is the function of structure refactoring kinetics of water molecules and minimum energy required for critical clusters. Therefore, the induction time of nucleation is heavily dependent on the supercooling degree and other kinetic parameters.

Concerning the above steps of hydrate formation, Vysniauskas and Bishnoi [13] proposed the consumption rate of methane gas could express the hydrate formation rate, and corresponding semi-empirical equation (Eq. 2.26) is determined by the concentration of water monomer, hydrate crystal nucleus and methane molecules, and gas-liquid interface area.

$$r_{\mathrm{H}} = A a_{\mathrm{s}} \exp\left(-\frac{\Delta E_{\mathrm{a}}}{RT}\right) \exp\left(-\frac{a}{\Delta T^{b}}\right) P^{\gamma} \tag{2.26}$$

where r_{H} is the consumption rate of methane, $\mathrm{cm}^3/\mathrm{min}$; A shows the comprehensive pre-exponential constant, $\mathrm{cm}^3/(\mathrm{cm}^2\cdot\mathrm{min}\cdot\mathrm{bar}^{\gamma})$, generally taking $4.554 \times 10^{-26}\mathrm{cm}^3/(\mathrm{cm}^2\cdot\mathrm{min}\cdot\mathrm{bar}^{\gamma})$; ΔE_{a} stands for the activation energy, KJ/mol, taking 106.204 kJ/mol; R represents the gas constant, J/(mol·K), taking 8.314 J/(mol·K); T refers to the system temperature, K; P refers to the system pressure, KPa; ΔT stands for the subcooling degree, K, that is the difference of hydrate equilibrium temperature and system temperature; a, b, γ show the experimental constants, $a = 0.0778\ \mathrm{K}^{b}$, $b = 2.411$, $\gamma = 2.986$; a_{s} is the area of gas-liquid interface, cm^2.

Based on the Vysniauskas and Bishnoi model, crystallography and double-membrane theory, Englezos et al. [14] established the intrinsic kinetic model of hydrate in the system of pure methane, pure ethane, mixing methane and ethane. However, the assumption—hydrate growth rate remains constant—is out of step with reality, and the related experiments are carried out in one static reactor with constant pressure. Thus, the description of hydrate growth underflow is greatly limited. Afterwards, by introducing two parameters of intrinsic kinetic, Turner et al. [15] proposed a simplified kinetic model of hydrate, which believed the subcooling degree comprises the driving force for hydrate formation. However, an empirical adjustment parameter needs to be introduced to describe the formation rate of hydrate formed by dispersion droplets.

2.2.2.2 Heat Transfer Model

The heat release would be found in the process of hydrate formation, and conversely, the timeliness of heat loss would exert a significant effect on the hydrate growth. Hence, the heat transfer model of hydrate argues that the process of heat transfer is the key element in restricting the hydrate formation. Based on the equilibrium relation between the rate of heat release and heat loss during hydrate formation, Uchida et al. [16] established the heat transfer model of hydrate (shown in Eq. 2.27), which simplifies the temperature gradient on the surface of the hydrate layer and believes hydrate growth rate is directly proportional to the driving force.

$$\frac{v_{\mathrm{f}}}{\lambda} = \frac{1}{(L\rho_{\mathrm{h}} r_{\mathrm{c}})} \Delta T \tag{2.27}$$

where L is the latent heat of hydrate formation, J/mol; r_{c} refers to the curvature radius of hydrate layer, m; ΔT refers to the driving force of system temperature, K; v_{f} shows the hydrate growth rate, m/s; λ stands for the heat transfer coefficient in the ambient environment of hydrate shell, W/(m·K); ρ_h represents the mole density of hydrate, mol/m^3.

By introducing the Nusselt number, Mori [17] believed the hydrate growth rate has an exponential relation to the driving force of temperature. Therefore, one simplified convective heat transfer model (shown in Eq. 2.28) was established according to the above heat transfer model, which holds that the hydrate growth is controlled by the heat transfer on the surface of hydrate crystal, and hydrate growth could be inhibited owing to the unresponsive heat loss.

$$\nu_f \delta = C \Delta T^{\frac{3}{2}} \tag{2.28}$$

$$C = \frac{\pi A}{4} \frac{1}{\rho_h \Delta h_h} \left(\rho_w c_{p,w} k_w^{\frac{2}{3}} + \rho_g c_{p,g} k_g^{\frac{2}{3}} \right) \tag{2.29}$$

where ν_f shows the hydrate growth rate, m/s; δ refers to the hydrate thickness, m; ΔT is the subcooling degree, K; C represents one parameter, $m^2/(K^{\frac{2}{3}} \cdot s)$; calculated through the Eq. 2.29; A stands for the dimensionless parameter; Δh_h is the latent heat of hydrate formation, J/kg; $c_{p,w}$, $c_{p,g}$ show the heat capacity of water, guest, J/(kg·K); ρ_h, ρ_w, ρ_g refer to the respective density of hydrate, water, and guest, kg/m^3; k_w, k_g represent the thermal diffusion coefficient of water and guest molecules, m^2/s.

Mochizuki and Mori [18] used the comprehensive boundary layer system (flat and hemispherical boundary layer) to investigate the heat transfer of the hydrate shell, and builded the involved model shown in Eqs. 2.30 and 2.31. It is assumed for the model that the diffusion growth of the hydrate layer was multidirectional and uniform.

$$\rho_h \delta \Delta h_h \nu_f = \int_0^{\delta} \left(\lambda_h \left. \frac{\partial T}{\partial x} \right|_{x=x_h^-} - \lambda_w \left. \frac{\partial T}{\partial x} \right|_{x=x_h^+} \right) dy \tag{2.30}$$

$$\rho_h \delta \Delta h_h \nu_f = \int_{-\frac{\pi}{2}}^{\frac{\pi}{2}} \left(\lambda_h \left. \frac{\partial T}{\partial r} \right|_{r=r_h^-} - \lambda_w \left. \frac{\partial T}{\partial r} \right|_{r=r_h^+} \right) dy \tag{2.31}$$

where Δh_h shows the latent heat of hydrate formation, J/kg; ν_f represents the hydrate growth rate along the boundary of water and guest molecules, m/s; T is the temperature, K; r_h stands for the initial boundary layer radius of hydrate shell, m; x_h stands for the initial position of hydrate layer formed in the x-axis direction, m; δ is the hydrate layer thickness, m; ρ_h is the hydrate density, kg/m^3; λ_h, λ_w show the heat transfer coefficient of hydrate and water, W/(m·K); "+" refers to that the boundary layer is close to the hydrate; "–" refers to that the boundary layer is close to the water.

2.2.2.3 Mass Transfer Model

The mass transfer model of hydrate considers that the diffusion of gas molecules to hydrate formation interface determines the hydrate formation rate. By simplifying the kinetic model of hydrate, combining the gas-liquid equilibrium with the solid-liquid equilibrium on the interface of hydrate formation, Skovborg and Rasmussen [19] proposed the following mass transfer model of hydrate.

$$\frac{dn}{dt} = k_L A_{(g-l)} c_{w0} x_b x_{int} \tag{2.32}$$

where $A_{(g-l)}$ is the area of the air-liquid interface, m^2; c_{w0} refers to the initial concentration of water molecules, mol/m^3; k_L stands for the mass transfer coefficient of liquid layer, m/s; n represents the mole of gas molecules consumed by hydrate growth, mol; t represents the time, s; x_{int} shows the mole fraction of gas under interfacial gas-liquid phase equilibrium; x_b shows the mole fraction of gas under the solid-liquid phase equilibrium.

Yapa et al. [20] proposed the mass transfer model of hydrate formed on the bubble surface, which believed that the concentration gradient of gas would strengthen the transportation of gas molecules to the hydrate formation surface and thus continuous hydrates would be formed.

$$\frac{d}{dr}\left(r^2 \frac{dC}{dr}\right) = 0, r_b \le r \le r_h \tag{2.33}$$

$$C(r_b) = C_0 \tag{2.34}$$

$$C(r_h) = C_i \tag{2.35}$$

$$-D_g 4\pi r_h^2 \psi_s \left.\frac{dC}{dr}\right|_{r=r_h} = \frac{dn}{dt} \tag{2.36}$$

where C refers to the mole number of gas in unit hydrate shell, mol/m^3; C_0 refers to the C value in the solid-gas interface, mol/m^3; C_i refers to the change value with hydrate growth, mol/m^3; D_g is the effective diffusion coefficient, m/s^2; n shows the mole number of consumed gas, mol; r represents the radial distance of bubble, m; r_b stands for the radius of a bubble, m; r_h stands for the radius of hydrate shell, m; t is the time, s; ψ_s is the irregular dimension coefficient of a non-spherical bubble.

By regarding the hydrate layer as one thin plate with tiny micropores and simplifying micropores into the curved capillaries, Mori et al. [17] established the mass transfer model of hydrate shown in Eq. 2.37, which takes into account the impact of δ_D, internal morphology of hydrate layer and mass transfer coefficient.

$$\frac{\delta\tau^2}{r_c\varepsilon} = \frac{\sigma\cos\theta}{4\mu_{\mathrm{w(g)}}Nk_{\mathrm{gw}}}\frac{1-(1+n)x_{\mathrm{gs}}}{x_{\mathrm{gs}}-x_{\mathrm{g}\infty}} \tag{2.37}$$

where k_{gw} shows the mass transfer coefficient of guest molecules on side of hydrate layer water phase, m/s; N is the hydration number; r_c stands for the capillary diameter, m; x_{gs} represents the solubility of guest molecules in the water phase, mol/m^3; $x_{g\infty}$ represents the solubility of guest molecules in the water phase as δ tends to infinity, mol/m^3; δ is the hydrate layer thickness, m; ε is the void ratio of hydrate layer; θ is the interface contact angle of water and guest molecules, rad; $\mu_{w(g)}$ refers to the water dynamic viscosity of saturated guest molecules, Pa s; σ shows the interface tension of water and guest molecules, N/m; τ shows the curvature of the capillary.

2.2.3 Hydrate Formation in Gas Phase with Free Water and Its Rate

Hydrate formation rate could be clarified by combining the hydrate formation kinetics, multi-phase flow and heat transfer characteristics of wellbore/pipeline in a deepwater gas well. The flow pattern is prone to the annular-mist flow when a small amount of free water is located in the pipeline. In this flow pattern, the free water can exist in two forms: liquid film on pipewall and droplet in the gas core, so the hydrate converted from droplets condensed in gas core and liquid film condensed on pipewall would be found. Furthermore, the formation process of hydrate formed by droplet in gas core follows the direction from the surface to the inside of the droplet until the droplet is completely converted to hydrate particles. Concerning the above principles, Wang et al. and Sun et al. [21–23] established a hydrate formation model in the gas phase with free water. The formation mechanism of hydrate converted from droplets condensed in gas core and liquid film condensed on pipewall is shown in Figs. 2.5 and 2.6.

At low flow velocity, most of the free water mainly exists on the pipewall in form of the condensed liquid film, and a small amount of free water exists in the gas phase in form of a droplet. However, at high flow velocity, the liquid film moves along the pipewall under the gas dragging force, then partly atomizes into droplets under the gas shearing force [24]. The droplet entrainment fraction (E) is the result when droplets in gas phase and liquid film atomization reach a balance, which represents the number of droplets in the gas phase and is defined as the proportion of droplet mass flow to the total liquid mass flow. E can be obtained by the semi-empirical equation proposed by Pan and Hanratty [25]:

$$\frac{E}{E_{\mathrm{m}}-E} = A_1\frac{DU_{\mathrm{g}}^2(\rho_{\mathrm{g}}\rho_{\mathrm{l}})^{\frac{1}{2}}}{\sigma} \tag{2.38}$$

$$E_{\mathrm{m}} = 1-\frac{W_{\mathrm{lfc}}}{W_{\mathrm{l}}} \tag{2.39}$$

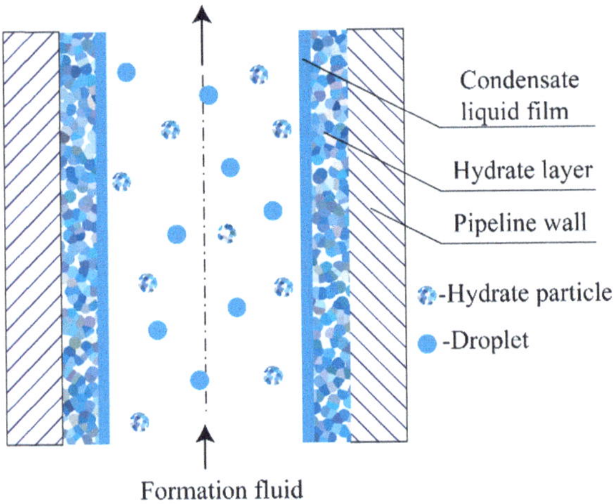

Fig. 2.5 Formation of hydrate converted from droplets condensed in gas core and liquid film condensed on pipewall in annular-mist flow

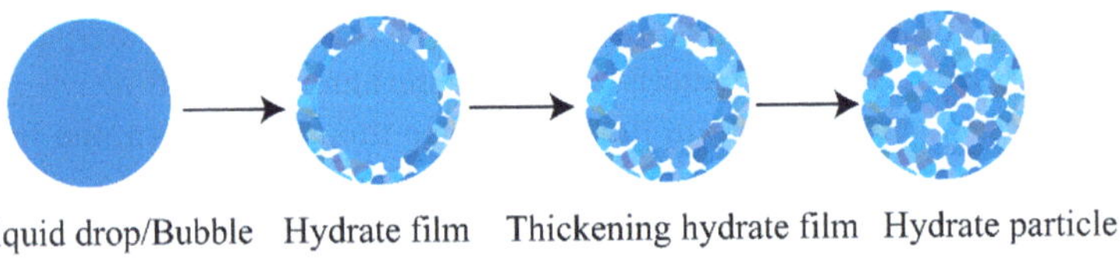

Fig. 2.6 Formation process of hydrate particle formed by droplet

$$W_{\mathrm{lfc}} = \pi D \Gamma_{\mathrm{c}} \tag{2.40}$$

$$\Gamma_{\mathrm{c}} = \frac{Re\mu_{\mathrm{l}}}{4} \tag{2.41}$$

$$Re = 7.3(\log_{10}\omega)^3 + 44.2(\log_{10}\omega)^2 - 263(\log_{10}\omega) + 439 \tag{2.42}$$

$$\omega = \frac{\mu_{\mathrm{l}}}{\mu_{\mathrm{g}}}\sqrt{\frac{\rho_{\mathrm{g}}}{\rho_{\mathrm{l}}}} \tag{2.43}$$

where σ shows the surface tension, mN/m; ρ_{g}, ρ_{l} refer to the density of gas and liquid, kg/m^3; D represents the internal diameter of the pipe, m; U_{g} is the gas flow rate, m/s; E_{m} stands for the maximum ratio of droplets entrainment in the gas phase, dimensionless and obtained by Eq. 2.39; W_{lfc} refers to the critical flow of the liquid film atomization on pipewall, kg/s, which is mainly impacted by the fluid property and flow velocity of the liquid film, and is calculated by Eq. 2.40 proposed by Andreussi et al. [26], Pan and Hanratty [25]; Γ_c shows the critical liquid film flow per length, kg/s; W_l refers to the liquid flow, kg/s.

Pan and Hanratty obtained the A_1 value of 8.8×10^{-5} under low pressure through one loop experiment. Lorenzo et al. [27] and Mantilla et al. [28] obtained the A_1 value of 3.6×10^{-5} by fitting the data of E under different flow velocities in high-pressure flow loop.

In annular-mist flow, only the surface of the liquid film is placed in contact with the gas phase, while the droplet in the gas core makes the most contact with the gas phase. Therefore, The formation rate of hydrate formed by droplet is greater than that by liquid film [21]. Aman et al. [29] found that the total surface area of droplets equals 4.75 times that of liquid film on pipewall under certain experimental conditions, which seemed to confirm the higher formation rate of hydrate formed by droplet. The gas-liquid contact area in unit control body (dl) could be expressed as [30]:

$$A_s = A_d + A_f \tag{2.44}$$

$$A_d = \frac{3\pi}{2} \cdot \frac{EQ_l}{SQ_g} \cdot \frac{D^2 \cdot dl}{d_{32}} \tag{2.45}$$

$$A_f = \pi(D - 2h)dl \tag{2.46}$$

where A_d shows the surface area of droplet, m^2; A_f shows the surface area of the liquid film, m^2; D is the inner diameter of the pipe, m; h represents the liquid film thickness, m; E refers to the droplet entrainment fraction; S is the ratio between the velocity of droplets and gas; Q_g, Q_l stand for the respective volumotric flow rates of gas and liquid, m^3/s; d_{32} shows the mean diameter of the entrained droplets, m;

The greater flow velocity of gas means the greater total surface area of droplets. Aman et al. [29] found the E in gas phase reached 5 and 22, and the droplet sizes were 54 and 20 μm under the respective gas flow velocity of 4.6 and 8.7 m/s, which caused that the difference on the total surface area of both droplets was highly up to eight times. The surface area of droplets in the control body with a length of dl could be calculated by Eq. 2.45 proposed by Aman et al., showing the surface area of droplets is affected by the numbers and sizes of droplets. Likewise, the surface area of the liquid film could be calculated by Eq. 2.46 [27], showing the impact of pipe inner diameter and liquid film thickness on the surface area of the liquid film.

Currently, the recognized models for hydrate formation rate mainly include the intrinsic kinetic model established by Turner et al. [31] and mass transfer model established by Skovborg and Rasmussen [19]. By comparing the measurement and calculation results of both models in annular-mist flow, Lorenzo et al. [27] found the calculation results in the kinetics model bringing into closer alignment with the experimental results. Hence, the hydrate formation rate could be expressed by the consumption rate of gas:

$$R_g = \frac{1}{M_g}\mu K_1 \exp\left(\frac{K_2}{T_s}\right) A_s T_{sub} \tag{2.47}$$

where M_g shows the mean molar mass of gas, g/mol; μ stands for one coefficient representing the transfer strength of mass and heat, dimensionless; taking 1/500 in the system dominated by the oil phase, taking 0.5 in the system dominated by gas phase [27]; K_1, K_2 refer to the kinetic parameter; taking 2.608×10^{16} kg $K/(m^2 \cdot s)$ as K_1 and −13600 K as K_2; T_s is the system temperature, K; T_{sub} is the thermodynamic subcooling degree, K.

During hydrate formation, the free water from the droplets and liquid film is consumed, which lead to the reduction of fluid flow and E. Accordingly, the distribution changes of the droplet and liquid film would affect the rate of hydrate formation and deposition, and the final distribution of δ. Conversely, this change of δ would feedback into the hydrate formation. Apparently, the calculation of free water consumption is critical for analyzing the rate of hydrate formation and deposition, and the variation of effective inner diameter.

Hydrate formation rate or free water consumption is different at different time and positions of wellbore/pipeline. Generally, the wellbore/pipeline is divided into multiple-unit control bodies to conduct the calculation of the fractional step of free water consumption. The schematic diagram of gas-liquid-solid three-phase flow with hydrate formation and deposition in dl is shown in Fig. 2.7 [32]. Supposing the distribution of temperature and pressure in dl is uniform, free water consumption in time dt could be expressed as Eq. 2.48. Given the mass conservation, the variation of liquid flow in time dt could be expressed as Eq. 2.49 [24, 33].

$$\mathrm{d}m_{l,j}^{i} = \mathrm{d}m_{le,j}^{i} + \mathrm{d}m_{lf,j}^{i} \tag{2.48}$$

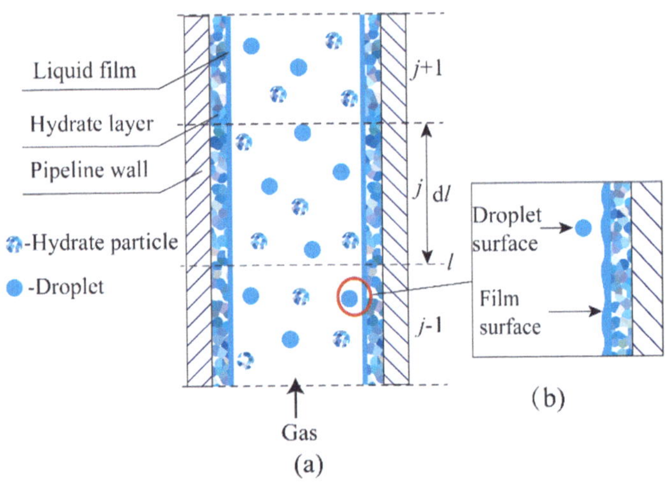

Fig. 2.7 Gas-liquid-solid three-phase flow with hydrate formation and deposition

$$W_{1,j+1}^{i+1} = W_{1,j}^{i} - \frac{dm_{1,j}^{i}}{dt} \tag{2.49}$$

where m_l represents the totally free water consumption in dl, kg; m_{le} represents the free water consumption from droplets, kg; m_{lf} represents the free water consumption from the liquid film, kg; W_l shows the liquid flow, kg/s; the subscript j indicates the number of unit control bodies in wellbore/pipeline; the superscript i indicates the time, s.

2.2.4 Hydrate Formation in Gas Phase Without Free Water and Its Rate

To prevent hydrate formation and pipe corrosion, gas is often dehydrated before entering the deepwater gathering pipeline. However, the processed water-saturated gas systems still contain some free vapor and tiny insoluble particles, which is mainly derived from mud and cuttings in drilling, ceramsite sand used for hydraulic fracturing of the reservoir, mixed hard lumps formed by the cuttings, dust, water and light hydrocarbons, etc. In one respect, with the continuous heat transfer between pipe fluid and low-temperature seawater, the saturated gas will condense into small droplets on the surface of suspended particles when the fluid temperature is lower than the gas dew point. For another, the temperature on pipewall is close to the low-temperature seawater, and the saturated water concentration of fluid near the pipewall is much lower than that of gas. So, the liquid film would be condensed on pipewall under the saturation concentration difference. Furthermore, the droplet condensed in gas core and condensed on pipewall will be converted and generated the hydrate instantly under the hydrate phase equilibrium state, which is shown in Fig. 2.8. Based

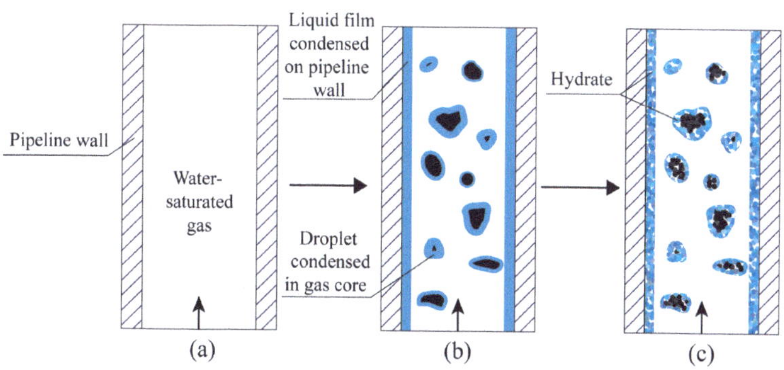

Fig. 2.8 Diagram of hydrate formation in water-saturated gas systems, **a** water-saturated gas single-phase flow, **b** free water condensed in gas core and on pipewall, **c** Formation of hydrate generated from droplets and liquid film

on the principles above, Zhang et al. [34] established one hydrate formation model in gas phase without free water.

2.2.4.1 Formation Mechanism of Hydrate Formed by Liquid Film

(1) Condensate Mechanism of Liquid Film

The condensate mechanism of the liquid film and its distribution are the basis for studying the mechanism of hydrate formation and deposition in deepwater gas pipelines. Conventional theory holds that the condensation phenomenon will be observed after water vapor contacts the cold surface of the pipeline when the system temperature is lower than the saturation temperature. Ordinarily, the whole process is divided into two ways: the rupture of the liquid film and condensation nucleation. Jakob [35] believed that a layer of liquid film on the cold wall firstly condensed out of the water vapor, then gradually thickened to reach one critical value, and ruptured with continuous diffusion of vapor to liquid film. After that, the liquid film is shrunk into multiple droplets under surface tension. Majumdar and Mezic [36] and Haraguchi et al. [37] corroborated the thin liquid film by observation during condensation of water vapor. Tammann and Boehme [38] first proposed the hypothesis of droplets condensation nucleation, which considered that the activation degree of vapor condensation in the different position of pipewall is varied. So, the vapor firstly is condensed into nucleation in the position of higher activation degree, then condensed droplets appear with the diffusion of vapor. Afterwards, the continuous growth of droplets is likely to make it merge and expand.

The condensate mechanism of fluid film is shown in Fig. 2.9 [34]. In the deepwater environment, the large temperature difference between the fluid in the pipe and the ambient environment would cause the dramatic changes of the thermal boundary layer at the radial direction near the pipewall. So, with the diffusion of water vapor in

Fig. 2.9 Condensate mechanism of liquid film

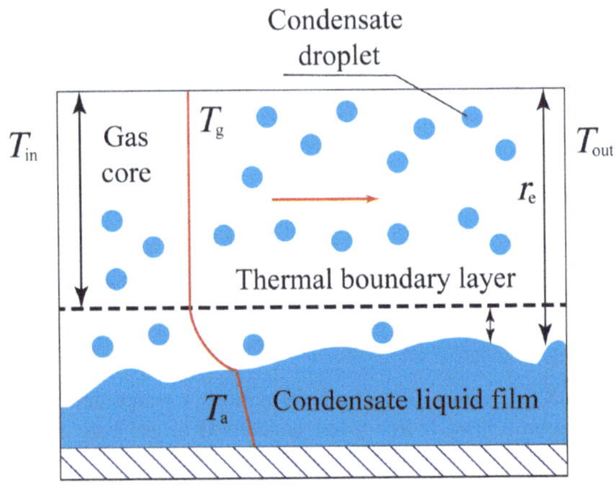

the gas core to the cold pipewall, the vapor would condense to form the liquid water under the difference of saturation concentration until it reaches saturation. Moreover, in the gas core beyond the thermal boundary layer, the little radial variations and large axial variations of temperature would make the small droplets on particles surface condense out of the saturated vapor. Notice that the liquid film and droplets are only for the interpretation of condensed free water. Actually, the hydrate formation rate is far greater than the condensation rate of free water, making it hard to observe it.

The liquid phase content in the system is mainly restricted by the content of saturated water in gas under different work conditions. The Mcketta–Wehe model of saturated water content is shown in Eq. 2.50, in which the coefficients of a_1, a_2 and a_3 are shown in Table 2.5.

The algorithm flowchart for the content of condensed free water is shown in Fig. 2.10. It can be seen that we need to calculate the water content of W_{i-1} and W_i under the same temperature and adjacent pressure, then obtain the saturated water

Table 2.5 Coefficients of a_1, a_2 and a_3

Pressure/Mpa	a_1	a_2	a_3
0.1	−30.0672	0.1634	−0.00017
0.2	−27.5786	0.1435	−0.00014
0.3	−27.8357	0.1425	−0.00014
0.4	−27.3193	0.1383	−0.00014
0.5	−26.2146	0.1309	−0.00013
0.6	−25.7488	0.1261	−0.00012
0.8	−27.2133	0.1334	−0.00013
1	−26.2406	0.1268	−0.00012
1.5	−26.129	0.1237	−0.00012
2	−24.5786	0.1133	−0.0001
3	−24.7653	0.1128	−0.0001
4	−24.7175	0.112	−0.0001
5	−26.8976	0.1232	−0.00012
6	−25.1163	0.1128	−0.0001
8	−26.0341	0.1172	−0.00011
10	−25.4407	0.1133	−0.0001
15	−22.6263	0.0973	−8.4E−05
20	−22.1364	0.0946	−8.2E−05
30	−20.4434	0.0851	−7E−05
40	−21.1259	0.0881	−7.5E−05
50	−20.2527	0.0834	−6.9E−05
60	−19.1174	0.0773	−6.2E−05
70	−20.5002	0.0845	−7.1E−05
100	−20.4974	0.0838	−7E−05

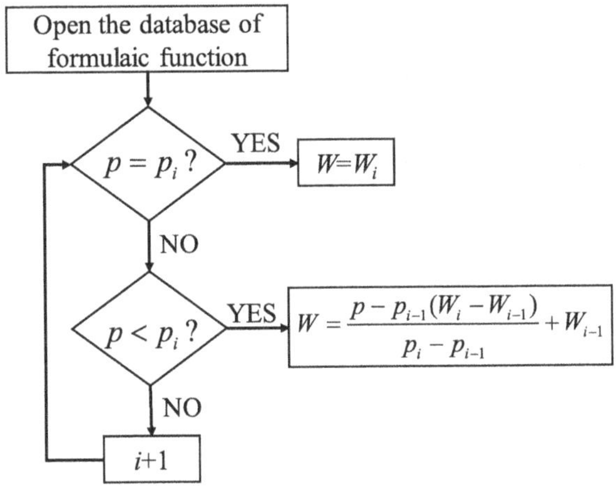

Fig. 2.10 Algorithm flowchart for the content of condensed free water

content in gas of W by difference calculation, finally correct the relative density of gas.

$$W = \alpha W_0 \tag{2.50}$$

$$W_0 = e^{a_1 + a_2 T + a_3 T^2} \tag{2.51}$$

$$\alpha = 0.001T + 0.369673d - 0.00142Td + 0.74217 \tag{2.52}$$

where W shows the saturated water content in gas, kg/m^3; W_0 shows the saturated water content in gas with the relative density of 0.6, kg/m^3, calculated by Eq. 2.51; α refers to the correction coefficient of gas relative density, calculated by Eq. 2.52; T is the absolute temperature, K; d stands for the relative density of gas.

The content of condensed liquid film could be represented by the difference between the content of saturated water in gas core and on pipewall with low temperature. Note that the hydrate layer should be regarded as the cold wall, on which the steam continuously condenses into the free water.

$$\Delta W_f = W_B - W_P \tag{2.53}$$

where W_B represents the content of saturated water in gas core, kg/m^3; W_P represents the content of saturated water on cold wall, kg/m^3.

(2) Formation of Hydrate Formed by Liquid Film

The hydrate formation mechanism in the deepwater pipeline is shown in Fig. 2.11 [34]. In deepwater pipeline dominated by gas phase, gas molecules and condensed droplets would continuously diffuse and transfer to the surface of the condensed liquid film and dynamic hydrate layer under low temperature.

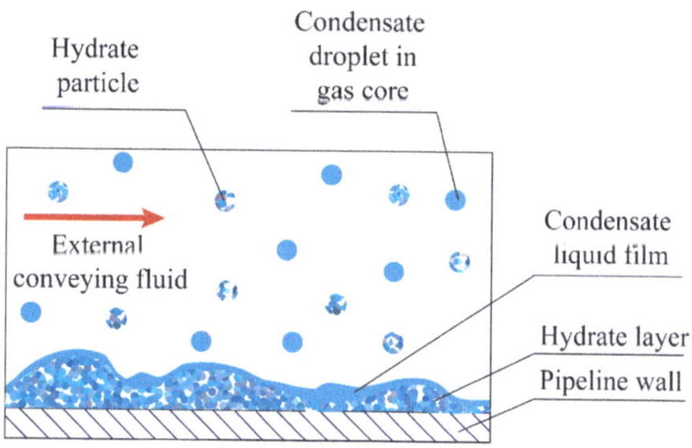

Fig. 2.11 Hydrate formation mechanism in deepwater pipeline

The distribution and content of the liquid phase in the pipeline are the keys to determine the hydrate formation rate. It is the free water in the pipeline dominated by gas-phase that restricts the amount of hydrate formation. Therefore, the consumption amount of free water could be used to represent the formation rate of hydrate converted from the condensed liquid film, which is shown by Eq. 2.54.

$$\frac{dm_{\mathrm{ld}}}{dt} = \alpha_2 2\pi r_{\mathrm{i}} h_{\mathrm{m}} \Delta W_{\mathrm{f}} ds \tag{2.54}$$

where dm_{ld} stands for the amount of hydrate converted from the condensed film, kg; h_{m} shows the coefficient of mass transfer, m/s, calculated by Eq. 2.56; ΔW_{f} refers to the amount of free water condensed on pipewall; r_{i} refers to the effective flow radius of the pipeline, m; α_2 represents the ratio of the relative molecular mass of hydrate to that of water in hydrate, considered as 1.15.

Based on the energy conservation of dl, the temperature T_i at the time i on pipewall is calculated as

$$2\pi r_{\mathrm{i}} h_{\mathrm{B}}(T_{\mathrm{B}} - T_i) = \frac{2\pi k_{\mathrm{h}}(T_i - T_{\mathrm{e}})}{\ln(r_{\mathrm{ti}}/r_{\mathrm{i}})} - 2\pi r_{\mathrm{i}} h_{\mathrm{m}} \Delta W_{\mathrm{f}} \Delta H_{\mathrm{f}} \tag{2.55}$$

$$Sh = 0.023 Re^{4/5} Sc^{1/3} = \frac{h_{\mathrm{m}} D}{D_{\mathrm{WC}}} \tag{2.56}$$

$$D_{\mathrm{WC}} = \frac{7.4 \times 10^{-8} (\varphi_{\mathrm{C}} M_{\mathrm{C}})^{1/2} T}{\mu_{\mathrm{C}} v_{\mathrm{W}}^{0.6}} \times 10^{-4} \tag{2.57}$$

where ΔH_{f} stands for the enthalpy of hydrate formation, J/kg; T_i shows the temperature at the time i on pipewall, °C; T_{e} shows the environment temperature, °C; T_{B} shows the fluid temperature, °C; k_{h} represents the thermal conductivity of hydrate, W/(m·K); r_{i} is the effective flow radius of the pipeline, m; r_{ti} is the initial inner diameter of the pipeline, m, the diameter distribution of pipeline is shown in Fig. 2.12; Sh refers to the Sherwood Number, dimensionless; Re refers to the Reynolds Number, dimensionless; Sc refers to the Prandtl Number, dimensionless; D_{WC} stands for

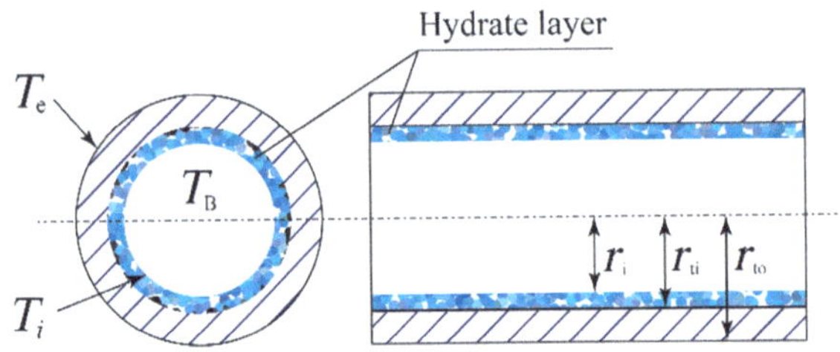

Fig. 2.12 Diameter distribution of pipeline

the mass transfer and diffusion coefficient of water molecules, m^2/s, calculated by Eq. 2.57; φ_C is the effective molecular mass during diffusion of water molecules, g; M_C is the molecular mass of fluid in the system, g/mol; μ_C represents the viscosity of the fluid in the system, mPa · s; ν_W refers the molar volume of water, cm^3/mol.

2.2.4.2 Formation Mechanism of Hydrate Formed by Droplet

(1) Heterogeneous Nucleation of Droplets

The heterogeneous nucleation theory has been widely applied for air pollution control and industrial production such as the artificial precipitation by sprinkling the condensation nucleus of dry ice and dust salt in the air, removal of industrial exhaust particles based on the principle of heterogeneous condensation.

In pure vapor system, the mutual collision and adhesion of vapor molecules could generate the microclusters, which would carry out homogeneous nucleation after it exceeds the critical radius. However, in the vapor system with suspended and micro-insoluble particles, the vapor molecules would first gather on the surface of the particles and then carry out heterogeneous nucleation after it exceeds the critical saturation. In 1958, Fletcher firstly proposed the heterogeneous nucleation theory of saturated vapor condensation on the surface of uniform and micro-insoluble spherical particles and found the particle radius is the main factor impacting the rate of droplet condensation nucleation. Later, Gorbunov and Hamilton [39, 40] verified the phenomenon of vapor condensation on the surface of micro-insoluble particles, and analyzed the influence of particle size and physicochemical properties of soluble components on the nucleation rate. Overall, the heterogeneous nucleation theory of vapor condensation provides one theoretical support for interpreting the condensation nucleation phenomenon on the surface of suspended particles in pipeline dominated by gas.

The particle's diameter and its surface properties have a great impact on the heterogeneous nucleation. Suppose that the free water would condense on particle's surface when the temperature and pressure reach the condition of liquid phase condensation. The total amount of condensed liquid phase and the number of condensed droplets in the gas core is shown in Eqs. 2.58 and 2.59, respectively [34].

$$\Delta W_T = W_{in} - W_{out} \tag{2.58}$$

$$\Delta W_{\mathrm{d}} = \Delta W_{\mathrm{T}} - \Delta W_{\mathrm{f}} \tag{2.59}$$

where ΔW_{T} shows the total amount of condensed liquid phase in dl, kg/m^3; W_{in} is the content of saturated water in the inlet of dl, kg/m^3; W_{out} is the content of saturated water in the outlet of dl, kg/m^3; ΔW_{f} refers to the amount of condensed liquid film in dl, kg/m^3; ΔW_{d} stands for the number of condensed droplets in dl, kg/m^3.

(2) Formation of Hydrate Converted from Droplet in Gas Core

Under low-temperature and high-pressure, the continuous diffusion of gas molecules to the droplet surface could contribute to forming a layer of hydrate shell on the interface, then a hydrate shell gradually grows inward to turn into the complete hydrate particles. This vigorous diffusion would cause the hydrate formation rate (10^{-10} kg/s) is much larger than the condensation rate of droplets (10^{-13} kg/s), so the impact of condensation rate of droplets on the hydrate formation could be neglected.

Suppose the droplets are spherical and its average size is controlled by the surface tension of droplets, the apparent velocity of gas-liquid phases and the content of gas-liquid phases. The hydrate formation rate could be calculated by Eq. 2.60.

$$\frac{\mathrm{d}m_{\mathrm{gf}}}{\mathrm{d}t} = \mu_{\mathrm{f}} A_{\mathrm{s}} K_1 M_{\mathrm{h}} \exp\left(-\frac{K_2}{T_{\mathrm{B}}}\right)(\Delta T_{\mathrm{sub}}) \tag{2.60}$$

$$\Delta T_{\mathrm{sub}} = T_{\mathrm{h}} - T_0 \tag{2.61}$$

$$t = 10^{2.1\cdot(\Delta T_{\mathrm{sub}} - 13.49) - 0.0225} \tag{2.62}$$

where dm_{gf} refers to the formation amount of hydrate particles in dl per unit time, kg; K_1, K_2 show the intrinsic kinetics constant, taking 2.608×10^{16} kg $\cdot$ K/(m$^2 \cdot$ s) and 13600 K for type II of hydrate, respectively; μ_{f} stands for the proportional coefficient representing the impact of mass and heat transfer on the hydrate formation rate, taking the range of 0.5–1 in the system dominated by gas proposed by Lorenzo et al. [27]; A_{s} is the contact area of gas and liquid phase in the gas core, m^2; M_{h} represents the molar mass of hydrate, kg/mol; ΔT_{sub} is the subcooling degree, K; the subcooling degree and corresponding induction time are calculated by Eqs. 2.61 and 2.62; T_{B} is the fluid temperature, K; T_{h} is the phase equilibrium temperature of hydrate, K; T_0 is the fluid temperature in pipeline, K; t refers to the induction time, min.

2.2.4.3 Solution and Validation of the Model

The formation rate of hydrate formed by liquid film and droplets is affected by the dynamic distribution of temperature and pressure, gas flow rate and the inner diameter of the pipe, and it is necessary to carry out the numerical solution. The solution steps are as follows:

(1) Divide the pipeline into multiple-unit control bodies with a length of ds, and set the shortest interval of dt. Suppose that the hydrate formation rate, gas flow rate, droplet concentration, liquid film thickness and pipeline inner diameter in dl are distributed evenly.
(2) Determine the initial parameters of unit control body j at the time i such as the gas flow rate $U_{g,j}^{i}$ pipeline inner diameter $r_{i,j}^{i}$, temperature $T_{B,j}^{i}$ and pressure p_{j}^{i}.
(3) Calculate the formation rates of hydrate formed by liquid film and droplets of $\frac{dm_{ld}}{dt}$ and $\frac{dm_{gf}}{dt}$ in dl according to the dynamic distribution of temperature and pressure.
(4) Calculate the parameters such as gas flow rate $U_{g,j+1}^{i+1}$, pipeline inner diameter $r_{i,j+1}^{i+1}$, temperature $T_{B,j+1}^{i+1}$ and pressure p_{j+1}^{i+1} of unit control body j + 1 at time i + 1 in view of the calculation result of unit control body j at the time i. Obtain the time-space distribution of hydrate formation rate by iteration.

Base on flow loop experiment, the hydrate formation rate model is verified by Rao et al. [41, 42]. Experimental conditions such as the pipe-in-pipe structure of 1 in × 1 in and 1/8 in, the methane gas under 22 °C filling two pipes, the volumetric flow rate of a pump of 6 L/s, cooling liquid under 1 °C filling the internal pipe of stainless steel are set up. The diagram of flow loop experiment is shown in Fig. 2.13.

The comparison results of model calculation and experimental data obtained by Rao, etc. is shown in Fig. 2.14. The average error of 11.4% between the two indicates

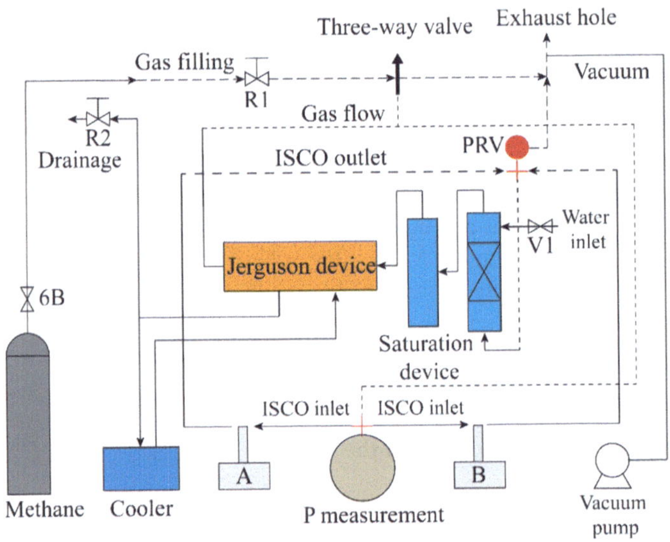

Fig. 2.13 Diagram of loop experiment conducted by Rao et al.

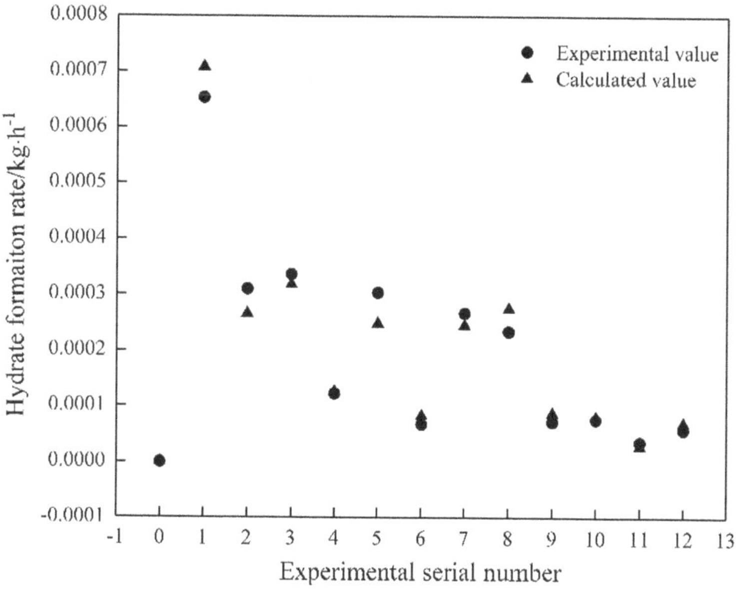

Fig. 2.14 Comparison of calculation result and literature data

the model calculation coincides with the experimental result to a great extent. Therefore, the hydrate formation rate without free water could be calculated accurately by this model.

2.3 Decomposition of NGH

Hydrate decomposition is essential for removing the hydrate blockage in pipeline, which involves the gas-liquid-solid three phases and is an endothermic reaction. Therefore, we need to provide sufficient heat to destroy the hydrogen bond in water cages and the Van der Waals force between guest and water molecules. Similar to the hydrate formation kinetics, hydrate decomposition kinetics are also controlled by intrinsic kinetics, mass transfer, and heat transfer.

If the hydrate decomposition is caused by the concentration difference of guest molecules between the hydrate surface and the surrounding solution, the process of hydrate decomposition would be limited by mass transfer. By obtaining the constant temperature and pressure at seabed to eliminate the restriction of heat transfer, Rehder et al. [43] measured the decomposition rates of CH_4 hydrate and CO_2 hydrate at the seabed, showing that the decomposition rate of CO_2 hydrate is significantly higher than that of CH_4 hydrate due to the solubility difference of gas.

The hydrate decomposition process also depends on the heat supply, and series experiments and simulations have shown the heat transfer model could predict the hydrate decomposition with high experimental precision. Sloan et al. established the heat transfer model to predict the hydrate decomposition time by depressurization and electric heating method, which described the heat transfer characteristic by Fourier method and its applicability was verified by the experiments and field data.

The hydrate decomposition above and below the freezing point (0 °C) has different features, so their decomposition mechanism and mathematical model shall be discussed separately.

2.3.1 Hydrate Decomposition Above Freezing Point

2.3.1.1 Mechanism and Model of Hydrate Decomposition by Thermal Stimulation

Ullerich et al. [44] and Selim and Sloan [45] proposed the heat transfer model to describe the decomposition process of methane hydrate based on the heat transfer law in the one-dimensional semi-infinite flat wall. Assuming that the water produced by hydrate decomposition is carried by gas and is away from the hydrate surface, the hydrate decomposition would be deemed as the ablation of moving interface. Kamath et al. [46, 47] considered the hydrate decomposition is controlled by heat transfer in the interface of the liquid film, which is strikingly similar to the nucleate boiling of fluid. Hence, the hydrate decomposition model is proposed as below:

$$\frac{m_{\mathrm{H}}}{\phi_{\mathrm{H}} A} = 6.464 \times 10^{-4} (\Delta T)^{2.05} \tag{2.63}$$

where m_{H} refers to the stable decomposition rate of hydrate, mol/h; ϕ_{H} is the volume fraction of hydrate, %; A stands for the interface surface area of hydrate layer and fluid, cm^2; ΔT represents the interface temperature difference of fluid and hydrate, °C.

Selim and Sloan [45] proposed the decomposition model of methane hydrate shown in Eq. 2.64.

$$X^* = \frac{St}{1+St}\left(\tau - \frac{1}{St}\right) \tag{2.64}$$

$$\tau = \frac{5 q_{\mathrm{s}}^2 t}{4 \rho_{\mathrm{H}} \lambda k (T_{\mathrm{s}} - T_{\mathrm{i}})} \tag{2.65}$$

$$X^* = \frac{5 q_{\mathrm{s}} X}{4k(T_{\mathrm{s}} - T)_{\mathrm{i}}} \tag{2.66}$$

$$St = \frac{\lambda}{C_{\mathrm{p}}(T_{\mathrm{s}} - T_{\mathrm{i}})} \tag{2.67}$$

where X^* refers to the position of hydrate decomposition interface, m; t is the time, s; T_s is the equilibrium temperature under system pressure, K; T_i is the initial temperature in the system, K; q_s shows the heat flux of hydrate decomposition surface, kW/m^2; k represents the heat conductivity coefficient of hydrate, 0.00039 kW/(m·K); ρ_H is the molar density of hydrate, $7.04 kmol/m^3$; λ shows the decomposition heat of hydrate, 3.31×10^5 kJ/kmol; St is the constant, calculated by Eq. 2.67.

2.3.1.2 Mechanism and Model of Hydrate Decomposition by Depressurization

Kim et al. [48] held that hydrate decomposition is a kinetic process of ignoring the mass transfer, which included the lattice rupture of hydrate particles, shrinkage of hydrate particles, desorption and escape of guest molecules. Base on this, Clarke and Bishnoi [49, 50] established the decomposition kinetics model of methane and ethane hydrate. Similarly, Sun et al. [51, 52] believed that the hydrate decomposition above freezing point includes two processes: water cage rupture of hydrate particle, desorption and escape of gas molecules, but the impact of hydrate decomposition area on its decomposition rate must be taken into account.

Kim et al. [48] proposed one hydrate decomposition model shown in Eq. 2.68 according to the following hypothesis. Suppose that the hydrate particles have the same volume and size; the number of hydrate particles is independent on reaction time; the mass and heat transfer resistance from gas to the surface of hydrate particle are neglected at high agitation rate.

$$-\frac{dn_H}{dt} = K_d A_s \left(f_{eq} - f\right) \tag{2.68}$$

where $-\frac{dn_H}{dt}$ stands for the hydrate decomposition rate; K_d is the intrinsic constant of hydrate decomposition rate, $mol/(MPa \cdot s \cdot m^2)$; A_s shows the total surface area of methane hydrate decomposition, m^2; f_{eq} represents the fugacity of methane gas under the condition of three-phase equilibrium, MPa; f represents the fugacity of methane gas in the experiment, MPa.

According to the experimental data, the activation energy of methane hydrate decomposition is 78.3 kJ/mol, which is similar to the methane hydrate formation heat of 62.8 kJ/mol [53], and the intrinsic constant of methane hydrate decomposition rate is 1.24×10^5 $mol/(Pa \cdot s \cdot m^2)$.

Based on Kim's model, Jamaluddin et al. [54] proposed the hydrate decomposition kinetic model considering the mass transfer and heat transfer, and considered that the surface roughness ψ has a small effect on hydrate decomposition rate under the small activation energy of hydrate decomposition ($E/R = 7553$ K), but a significant effect under the large activation energy ($E/R = 9400$ K). Besides, the hydrate decomposition is mainly controlled by heat transfer when $\psi > 64$, and with the variation of system pressure, the control mechanism of hydrate decomposition may transform from the single heat transfer to the combination of heat transfer and intrinsic kinetics of hydrate decomposition.

2.3.2 *Hydrate Decomposition Below Freezing Point*

Handa [55] proposed the two-step mechanism of hydrate decomposition below the freezing point. the surface of the hydrate layer starts to rapid hydrate decomposition, then the ice layer form by decomposed water covers the surface of hydrate particles to prevent its further decomposition, which is called the self-sealed effect of the hydrate. Takeya et al. [56–58] verified the ice layer on the surface of hydrate particles by X-ray diffraction, and found the surface structure of hydrate changed with the temperature. Sun et al. [51, 52] believed that below the system temperature of 0 °C, the undecomposed hydrate particles would be wrapped by the ice layer generated from the free water that is produced by hydrate decomposition, and the decomposition gas would diffuse out along the gap in hydrates and ice. With the continuous decomposition of hydrate, the ice layer would gradually thicken, or the hydrate layer would gradually be thin. Therefore, the hydrate decomposition below 0 °C could be regarded as the moving process of the ice-hydrate interface. Yan and Liu [59] proposed that the decomposition process of methane hydrate in the porous media of activated carbon includes two stages: the desorption and escape of methane molecules after the hydrate lattice rupture; the diffusion of methane molecules in liquid film or ice layer. Lin [60] described the ice layer as one gradually thickened and porous spherical shell, and proposed the decomposition process of single hydrate particle is also divided into the above two stages.

Currently, the two-step mechanism of hydrate decomposition with an ice layer covering, proposed by Handa [55], is widely recognized. Based on this, many scholars have established the mathematical models of hydrate decomposition. Some representative dynamics models are introduced below.

2.3.2.1 Spherical Model

Based on the theory of ice layer covering, Takeya et al. [56–58] established the decomposition kinetics model of single spherical hydrate particle controlled by diffusion as follows:

$$3\left(1-R^2\right)+2\left(R^3-1\right)=\frac{6D}{r_{\mathrm{h,0}}^2}\left[\frac{C_{\mathrm{d}}(T)-C_{\mathrm{a}}}{C_0-C_{\mathrm{a}}}\right] \tag{2.69}$$

$$R=r_{\mathrm{h}}/r_{\mathrm{h,0}} \tag{2.70}$$

where r_{h} shows the radius of hydrate particle, μm; $r_{\mathrm{h,0}}$ shows the initial outer diameter of hydrate particle, pm; T refers to the time of hydrate decomposition, s; D is the diffusion coefficient of hydrate in ice layer, m^2/s; $C_d(T)$ refers to the methane concentration in the gas phase under the temperature of T, mol/m^3; C_0 refers to the methane concentration in hydrate, mol/m^3; C_a refers to the methane concentration in air, mol/m^3.

According to the above model, the gas diffusion coefficient is $2.2 \times 10^{-11}\text{m}^2/\text{s}$ and $9.6 \times 10^{-12}\text{m}^2/\text{s}$ under the respective temperature of 189 and 168 K. The higher diffusion coefficient indicates that the methane gas would diffuse by the ice layer, the pore channels or boundaries between hydrate particles.

2.3.2.2 Model of Ice-Hydrate Interface Moving

In terms of the decomposition properties of methane and carbon dioxide hydrate, Sun et al. [51, 52] established a kinetics model of hydrate decomposition below the freezing point. Assume that the molar volume of ice is approximately equal to that of the hydrate. The mass conservation on hydrate surface (Eq. 2.71) could be computed by correlating the hydrate decomposition rate with the decrease of δ or the increase of ice layer thickness.

$$-\frac{\mathrm{d}n}{\mathrm{d}t} = -A_{\text{geo}}\rho_{\text{hyd}}\frac{\mathrm{d}S}{\mathrm{d}t} \tag{2.71}$$

where A_{geo} shows the geometric surface area of hydrate, m^2; S represents the thickness of the ice layer, m; ρ_{hyd} is the gas density in hydrate, kg/m^3; n stands for the mole number of gas component, dimensionless.

2.3.2.3 Porous Spherical Shell Model

Lin et al. [60] established a decomposition model of single hydrate particle controlled by interface chemical reaction and gas diffusion, and derived the total reaction model. Suppose that sensible heat from the hydrate sediment would be provided in time; the diameter and shape of all the hydrate particles maintain constant during hydrate decomposition. The differential equation controlled by interface chemical reaction and gas diffusion is shown in Eqs. 2.72 and 2.73 respectively.

$$\frac{\mathrm{d}r_c}{\mathrm{d}t} = k_d\left(C_g - C_{eq}\right)/\rho_s \tag{2.72}$$

$$\frac{\mathrm{d}r_c}{\mathrm{d}t} = \frac{D_\infty/\rho_s}{\left(1 - \frac{r_c}{r_0}\right)r_c}\left(C_0 - C_g\right) \tag{2.73}$$

where k_d shows the constant representing the hydrate decomposition rate, cm/s; r_c is the radius of undecomposed hydrate particle, cm; C_g stands for the gas concentration on the surface of hydrate particle, mol/cm^3; C_{eq} stands for the gas concentration under the condition of three-phase equilibrium, mol/cm^3; ρ_s is the gas concentration in hydrate, 0.0083 mol/cm^3; C_0 shows the gas concentration, mol/cm^3; r_0 represents the radius of hydrate particle at some point, cm; D_∞ is the diffusion coefficient, cm^2/s.

Research suggests that the hydrate decomposition rate is mainly controlled by interface chemical reaction in the initial stage, but by the gas diffusion in the middle and late stages. Therefore, the total hydrate decomposition rate model could be derived shown in Eq. 2.74, in which the adjustable parameters K and k_d could be obtained by fitting the experimental data with simplex optimization, and whose numerical solution could be calculated by the Euler dynamics iterative algorithm.

$$\frac{dx}{dt} = K\frac{C_0 - C_{eq}}{\left(\frac{1}{k_d} + \frac{r_0 x^{1/3}(1-x^{1/3})}{D}\right)x^{-2/3}} \tag{2.74}$$

2.4 Formation and Decomposition of Hydrate Converted from Moving Bubbles

Under low-temperature, the bubbles migration in wellbore would accelerate the phase transition between gas, liquid, and hydrate. During this process, the laws of interphase mass and heat transfer have a significant impact on the bubbles' motion characteristics and wellbore dynamics. Therefore, to predict the migration of gas phase in the wellbore, control the wellbore pressure and finally attain our goal of well flow assurance, it is necessary to carry out the quantitative analysis for the growth and renewal of hydrates generated on the surface of moving bubbles. Based on the above principles, Sun et al. [61] proposed the formation and decomposition model of moving bubbles hydrate.

2.4.1 *Mass Transfer Model of Moving Bubbles*

The mass transfer rate between phases on the surface of the moving bubbles is not only closely related to the non-uniform flow field, the boundary layer of mass transfer and the hydrate shell morphology, but it is a key factor to describe the gas-liquid-phase action and migration state in bubble flow [61, 62]. For the prediction of the mass transfer rate, it is necessary to quantitatively describe how the hydrates on the surface of moving bubbles grow and renew. As Fig. 2.15 shows, the sub-processes affecting the growth and renewal rate of hydrate shells mainly include gas dissolution, hydrate formation, hydrate dissolution and hydrate decomposition [62].

The hydrate shells grow and renew in the stability area of the hydrate. The hydrate formation and gas-liquid mass transfer on the surface of moving bubbles is shown

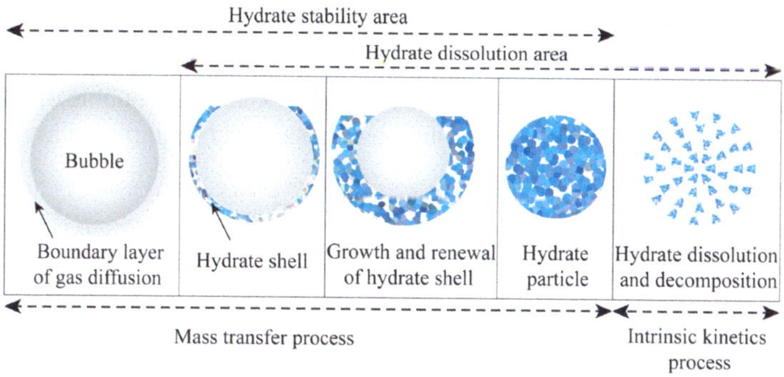

Fig. 2.15 Growth and renewal of hydrate shell on the surface of moving bubble

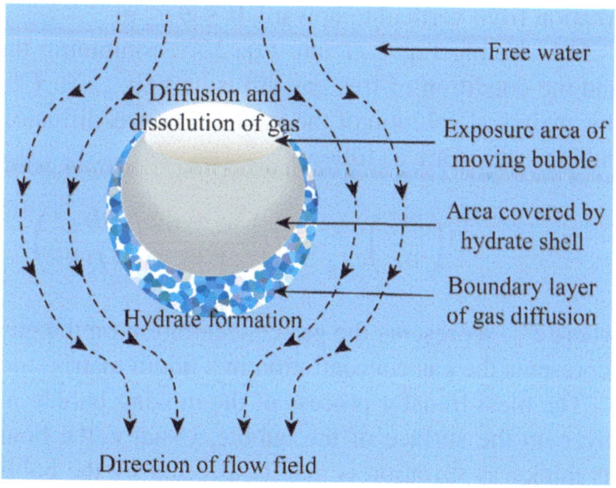

Fig. 2.16 Hydrate formation and gas-liquid mass transfer process on the surface of moving bubble

in Fig. 2.16 [63], which shows the free gas would pass through the bubble exposed area and be dissolved in the liquid phase. Meanwhile, the dissolved gas continues migrating to the surface of the hydrate shell to form the solid hydrate. Notice that the hydrate shell mainly gathers and covers the lower part of the bubble under the action of the flow field, boundary layer of gas diffusion and shear force.

Suppose that the hydrate shell partially covers the bubble surface; the liquid phase quickly reaches the saturated state of gas dissolution in the multi-phase flow; the split and coalescence of moving bubbles, the nucleation and shedding of hydrate crystal are not considered, whereas the hydrates dissolution needs to be taken into account. Therefore, the mass transfer model of moving bubbles in the liquid phase is shown in Eq. 2.75, which is mainly applied to the bubble flow.

$$v_r \frac{\partial C}{\partial r} + v_\theta \frac{1}{r} \frac{\partial C}{\partial \theta} = D\left(\frac{\partial^2 C}{\partial r^2} + \frac{2}{r} \frac{\partial C}{\partial r}\right) \tag{2.75}$$

where C shows the gas concentration, mol/m^3; D shows the gas diffusion coefficient, m^2/s; r is the radial distance of the moving bubble, m; ν_r stands for the radial velocity component, m/s; ν_θ stands for the angular velocity component, m/s.

Since the thickness of the boundary layer is far less than the bubble diameter, the velocity potential function of flow around bubbles could be simplified as:

$$\phi = -\frac{U_\infty}{2}\sin^2\theta\left(r^2 - \frac{3}{2}Rr + \frac{1}{2}\frac{R^3}{r}\right) \approx -\frac{3}{4}U_\infty \sin^2\theta(r-R)^2 \tag{2.76}$$

$$\nu_r = \frac{1}{r^2\sin\theta}\frac{\partial\phi}{\partial\theta},\ \nu_\theta = -\frac{1}{r\sin\theta}\frac{\partial\phi}{\partial r} \tag{2.77}$$

where R stands for the bubble radius, m; r is the radial distance of bubble, m; U_∞ refers to the velocity difference between gas and liquid phase in multi-phase flow, m/s; θ represents the exposure angle of the bubble, rad, that is the angle of clockwise rotation from vertical to upward, $0 \le \theta \le \pi$.

Substituting Eq. 2.77 into Eq. 2.75, combining the variable separation method and the condition of free gas diffusion ($C|_{r=R} = Cb_{\text{int}}$, $C|_{r=\infty} = 0$) could obtain the analytical solution of the model. The gas diffusion rate at different positions of bubble surface are as follows:

$$M_{\text{A}}(\theta) = D\left(\frac{\partial C}{\partial r}\right)_{r=R} \approx D\frac{C_{\text{int}} - C_{\text{b}}}{1.15}\left(\frac{3U_\infty}{4DR^2}\right)^{1/3}\frac{\sin\theta}{(\theta - \sin\theta\cos\theta)^{1/3}} \tag{2.78}$$

where C_{int} represents the gas concentration on the gas-liquid interface, mol/m^3; C_{b} represents the gas concentration in a liquid matrix, mol/m^3.

The mass transfer process of the moving bubble mainly occurs in the boundary layer on the surface of the bubble. Usually, the boundary layer is very thin, and its thickness variation is mainly affected by the bubble size and flow field around bubbles, but is irrelevant to the distribution of gas concentration. The thickness of mass transfer boundary layer (δ_c) is calculated by Eq. 2.79.

$$\delta_c = \frac{M_{\text{A}}(\theta)}{D(C_{\text{int}} - C_{\text{b}})} = 1.15\frac{(\theta - \sin\theta\cos\theta)^{1/3}}{\sin\theta}\left(\frac{4DR^2}{3U_\infty}\right)^{1/3} \tag{2.79}$$

The δ_{c} distribution around bubbles is uneven, which gradually increases with the exposure angle of the bubble. Thus, it is why hydrate shell first grows in the area with high dissolved gas, that is the lower part of bubble, then covers the surface of bubble. By now, the bubble covered by a layer hydrate shell could be simply called the "hydrate bubble". For the single "hydrate bubble" in the hydrate phase equilibrium region, the mole changes of "hydrate bubble" caused by gas dissolution and hydrate formation is

$$\Delta J = J_{\text{gL}} - J_{\text{sh}} \tag{2.80}$$

where J_{gL} shows the mass transfer rate between gas and liquid phase for a single bubble, mol/s; J_{sh} refers to the mass transfer rate between dissolved gas and hydrate for a single bubble, mol/s.

The mass transfer on the bubble surface between gas-liquid-hydrate phases is influenced by the covering pattern of the hydrate shell, and the gas directly contacts with the liquid phase in the upper exposure area ($\varphi \le \theta \le \pi$). The gas consumption rate caused by gas dissolution is shown in Eq. 2.81. However, in the coverage area of the hydrate shell ($\varphi \le \theta \le \pi$), the mass transfer between gas-liquid phases is limited. The gas consumption rate caused by hydrate formation is

$$J_{gL} = \int_0^{\varphi} 2\pi R^2 \sin\theta M_A(\theta) d\theta = (C_{gL} - C_b) 2\pi R^2 \frac{D}{1.15} \left(\frac{3U_\infty}{4DR^2}\right)^{1/3} \int_0^{\varphi} \frac{\sin^2\theta}{(\theta - \sin\theta\cos\theta)^{1/3}} d\theta \tag{2.81}$$

$$J_{sh} = A_h k_h (f_b - f_{eq}) = 2\pi R^2 (1 + \cos\theta) k_h (f_b - f_{eq}) \tag{2.82}$$

$$f_{eq} = f_{eq}(T, p_{eq}) \tag{2.83}$$

$$f_b = f_{sol}(p, T) = f^{sat} \exp\left(\int_{p^{sat}}^{p} \frac{\overline{v_g} dp}{R_0 T}\right) \tag{2.84}$$

where C_{gL} refers to the gas solubility in the liquid phase under gas-liquid phases equilibrium, mol/m^3; A_h represents the coverage area of hydrate shell, m^2; k_h is the constant representing the hydrate formation on bubble surface, mol/(m^2·Pa); $f_b - f_{eq}$ show the driving force of hydrate formation, that is the fugacity difference between dissolved gas and gas-liquid-solid phases equilibrium; p^{sat} stands for the saturation pressure of the gas, Pa; f^{sat} stands for the gas fugacity under saturation pressure, Pa; $\overline{v_g}$ shows the partial molar volume of gas in solution, mol/m^3; R_0 is the gas constant, J/(K·mol); p_{eq} refers to the hydrate phase equilibrium pressure, Pa; T is the system temperature, K.

Gas concentration in the liquid phase would reach the saturated state under a quasi-steady state. At this moment, the fugacity of dissolved gas is the function of temperature and pressure shown in Eq. 2.84, and the gas consumption rate caused by gas dissolution is equal to the rate caused by hydrate formation ($\Delta J = 0$). Substituting Eqs. 2.81 and 2.82 into Eq. 2.80 could obtain the Eq. 2.85, and its solution could get the exposure area of bubble and rate of interphase mass transfer under quasi-steady state. Also, the morphology of the "hydrate bubble" is shown in Eq. 5.86, in which the bubble radius is shown in Eq. 5.87.

$$\int_0^{\varphi} \frac{\sin^2\theta}{(\theta - \sin\theta\cos\theta)^{1/3}} d\theta = \frac{f_{sol} - f_{eq}}{C_{gL} - C_b} \frac{1.15 k_h}{D} \left(\frac{4DR^2}{3U_\infty}\right)^{1/3} \tag{2.85}$$

$$\delta = R - R\left[1 - \left(\frac{\alpha_{\mathrm{h}}}{\alpha_{\mathrm{g}} + \alpha_{\mathrm{h}}}\right)\frac{2}{(1 - \cos\theta)}\right]^{1/3} \tag{2.86}$$

$$R = \left[\frac{3}{4N\,\pi}(\alpha_{\mathrm{g}} + \alpha_{\mathrm{h}})\right]^{1/3} \tag{2.87}$$

where δ stands for the hydrate shell thickness on the bubble surface, m; $\alpha_{\mathrm{g}}, \alpha_{\mathrm{h}}$ represent the volume fraction of gas and hydrate in multi-phase flow, %.

For hydrate shell formation, the condition for hydrate formation on bubble surface $\left(f_{\mathrm{eq}} \le f_{\mathrm{b}}\right)$ shows that the hydrate shell starts to form when the volume fraction of dissolved gas exceeds a certain critical value. The lower subcooling degree suggests the smaller value of f_{eq} or smaller volume fraction of dissolved gas for hydrate formation, so the dissolved gas will be at the under-saturated state. When the dissolved gas concentration is smaller than its solubility ($f_{\mathrm{b}} \le f_{\mathrm{sol}}$), the gas consumption rate caused by hydrate formation is lower than the gas dissolution rate $\left(J_{\mathrm{sh}} \le J_{\mathrm{gl}}\right)$, leading to the gradual increase of dissolved gas concentration. Therefore, the dissolved gas will be at the hypersaturated state ($f_{\mathrm{sol}} \le f_{\mathrm{b}}$), and the gas consumption rate caused by hydrate formation will increase $\left(J_{\mathrm{gl}} \le J_{\mathrm{sh}}\right)$, resulting in the dissolved gas will stabilize its saturation state.

For the hydrate consumption, two models of hydrate consumption of hydrate dissolution and hydrate decomposition are found. The former mainly occurs at the under-saturated region of dissolved gas and is mainly influenced by mass transfer of gas. However, hydrate decomposition only occurs outside the hydrate phase equilibrium region due to the limitation on temperature and pressure. The hydrate consumption rate is calculated as

$$J_{\mathrm{hg}} = J_{\mathrm{diss}} + J_{\mathrm{decom}} \tag{2.88}$$

where $J_{\mathrm{diss}}, J_{\mathrm{decom}}$ refer to the dissolution and decomposition rate of hydrate respectively, mol/s. Generally, the hydrate decomposition rate is much larger than the former. Furthermore, in the hydrate phase equilibrium region, the hydrate decomposition rate could be neglected so that the hydrate consumption rate stands for its dissolution rate. Otherwise, outside the hydrate phase equilibrium region, the hydrate consumption rate is approximately equal to its decomposition rate.

In multiphase flow, the convective form is the main pattern of mass transfer as the "hydrate bubble" migrates to the under-saturated region of dissolved gas. Concerning the local coverage of hydrate shell on the bubble surface, the hydrate dissolution rate is

$$J_{\mathrm{diss}} = \frac{\rho_{\mathrm{h}}}{M_{\mathrm{h}}}\int_{\varphi}^{\pi} 2\pi R^2 \sin\theta \dot{R}\mathrm{d}\theta = \frac{\rho_{\mathrm{h}}}{M_{\mathrm{h}}}\int_{\varphi}^{\pi} 2\pi R^2 \sin\theta \frac{C_{\mathrm{sat}} - C_{\mathrm{b}}}{C_{\mathrm{cry}} - C_{\mathrm{sat}}}\frac{D}{\delta_{\mathrm{c}}}\mathrm{d}\theta \tag{2.89}$$

where C_{sat} represents the saturated gas concentration on the liquid-hydrate interface, mol/m^3; C_{b} represents the gas concentration in a liquid matrix, mol/m^3; C_{cry}

represents the gas concentration in hydrate, mol/m^3; $\dot{R}$ shows the variation rate of bubble radius, m/s; R shows the bubble radius, m; θ refers to the exposure angle of the bubble, rad; ρ_h is the hydrate density, kg/m^3; M_h is the molar mass of hydrate, kg/mol.

Substituting Eq. 2.79 into Eq. 2.89 could obtain the hydrate dissolution rate as follows:

$$J_{\text{diss}} = 1.739\pi R^2 D \frac{\rho_h}{M_h} \frac{C_{\text{sat}} - C_b}{C_{\text{cry}} - C_{\text{sat}}} \left(\frac{3U_\infty}{4DR^2} \right)^{1/3} \int_\varphi^\pi \frac{\sin^2\theta}{(\theta - \sin\theta\cos\theta)^{1/3}} d\theta \tag{2.90}$$

When the "hydrate bubble" migrates to the region outside hydrate phase equilibrium, hydrate shell will quickly decompose into hydrate particles, Its decomposition rate is expressed as

$$J_{\text{decom}} = K_d^0 \exp(-\Delta E / R_0 T) A_s (f_{eq} - f_b) \tag{2.91}$$

where K_d^0 refers to the constant representing the hydrate decomposition rate, $\text{mol}/(\text{m}^2 \cdot \text{Pa} \cdot \text{s})$; ΔE shows the activation energy of hydrate decomposition, J/mol; R_0 is the gas constant, J/(k mol); A_s represents the total surface area of hydrate decomposition, m^2.

2.4.2 *Interphase Mass Transfer Rate*

Equations 2.81, 2.82, 2.90, 2.91 refer to the mass transfer rate for gas dissolution on bubble surface, hydrate formation, hydrate dissolution, and hydrate decomposition respectively. Combining the rules of microscopic growth of hydrate shell, interphase mass transfer and microcosmic multi-phase flow in the wellbore, the interphase mass transfer model of "hydrate bubble" under the bubble flow can be obtained. The interphase mass transfer rate could be expressed by

$$\dot{m}_{ij} = \frac{(\alpha_g + \alpha_h) A}{4\pi R^3} J_{ij} M_i \tag{2.92}$$

where i, j stands for each phase including gas, liquid, and hydrate; M_i shows the molar mass of i phase, kg/mol; A represents the effective section area, m^2; α_g, α_h show the volume fraction of gas and hydrate in bubble flow, %; R shows the bubble radius, m.

2.4.3 Formation and Decomposition of Hydrate Converted from Moving Bubbles

In this section, by taking into account the flow field distribution and the non-uniform mass transfer boundary layer during the upward migration of bubbles, the established mass transfer model is applied to simulate the velocity field and radius of bubbles with initial bubble radius of 2 mm, and hydrate growth under the condition of bubble flow. The velocity potential of moving bubble and distribution of mass transfer boundary layer is shown in Fig. 2.17, illustrating that under the action of the flow field, δ_c gradually increases from the top to the bottom of bubbles during the upward migration of the bubble, which is caused by the non-uniform mass transfer rate around the moving bubble.

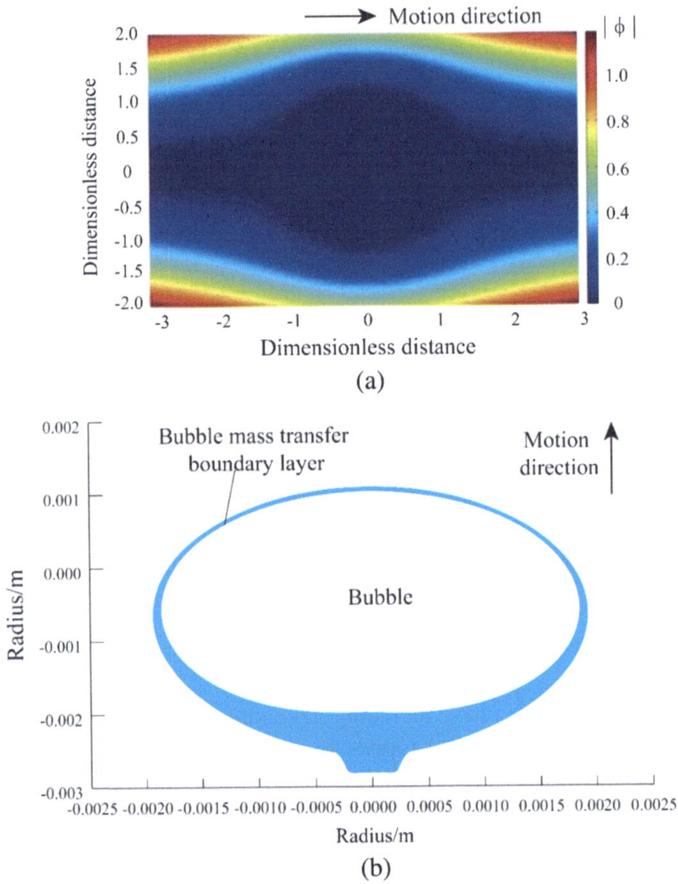

Fig. 2.17 Velocity potential of bubble (**a**) and distribution of mass transfer boundary layer (**b**)

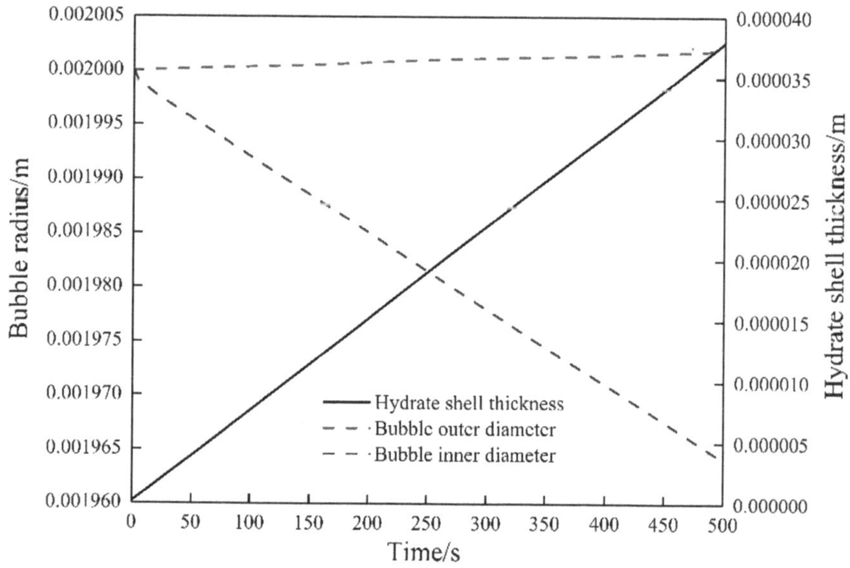

Fig. 2.18 Change of single bubble radius and hydrate shell thickness

The initial bubble pressure and subcooling degree are 10 MPa and 2.0 K respectively. The change of bubble radius and hydrate shell thickness under the saturation state of dissolved gas is shown in Fig. 2.18, illustrating the solid hydrate would form and its shell thickness around the moving bubble would gradually increase with time. That is to say, the bubble inner radius gradually decreases with the gas consumption, while the bubble outer radius slowly increases due to the larger hydrate volume comparing with the original gas volume.

References

1. Dendy SE (2003) Fundamental principles and applications of natural gas hydrates. Nature 426(6964):353–363
2. Taylor CJ, Miller KT, Koh CA et al (2007) Macroscopic investigation of hydrate film growth at the hydrocarbon/water interface. Chem Eng Sci 62(23):6524–6533
3. Sloan ED Jr, Koh CA (2007) Clathrate hydrates of natural gases. CRC Press, Boca Raton
4. van der Waals JH, Platteeuw JC (1958) Clathrate solutions. Adv Chem Phys 1–57
5. Liu P (2011) Study on prediction model of pipeline gas hydrate formation position. Xi'an Shiyou University
6. Ma Q, Chen G, Sun C (2008) Gas hydrate science and technology. Chemical Industry Press
7. Holder GD, Angert PF (1982) Simulation of gas production from a reservoir containing both gas hydrates and free natural gas. In: SPE annual technical conference and exhibition. Society of Petroleum Engineers
8. Kirchner MT, Boese R, Billups WE et al (2004) Gas hydrate single-crystal structure analyses. J Am Chem Soc 126(30):9407–9412

9. Parrish WR, Prausnitz JM (1972) Dissociation pressures of gas hydrates formed by gas mixtures. Ind Eng Chem Process Des Dev 1(11):26–35
10. Ng HJ, Robinson DB (1985) Hydrate formation in systems containing methane, ethane, propane, carbon dioxide or hydrogen sulfide in the presence of methanol. Fluid Phase Equilib 21(1):145–155
11. Katz DL (2013) Prediction of conditions of hydrate formation in natural gases. Trans Aime 160(1):140–149
12. Guang JC, Tian MG (1996) Thermodynamic modeling of hydrate formation based on new concepts. Fluid Phase Equilib 122(1–2):43–65
13. Vysniauskas A, Bishnoi PR (1983) A kinetic study of methane hydrate formation. Chem Eng Sci 38(7):1061–1072
14. Englezos P, Kalogerakis N, Dholabhai PD et al (1987) Kinetics of formation of methane and ethane gas hydrates. Chem Eng Sci 42(11):2647–2658
15. Turner D, Boxall J, Yang S et al (2005) Development of a hydrate kinetic model and its incorporation into the OLGA2000® transient multiphase flow simulator. In: 5th international conference on gas hydrates, Trondheim, Norway 12–16
16. Uchida T, Ebinuma T, Kawabata J et al (1999) Microscopic observations of formation processes of clathrate-hydrate films at an interface between water and carbon dioxide. J Cryst Growth 204(3):348–356
17. Mori YH (2001) Estimating the thickness of hydrate films from their lateral growth rates: application of a simplified heat transfer model. J Cryst Growth 223(1):206–212
18. Mochizuki T, Mori YH (2006) Clathrate-hydrate film growth along water/hydrate-former phase boundaries—numerical heat-transfer study. J Cryst Growth 290(2):642–652
19. Skovborg P, Rasmussen P (1994) A mass-transport limited model for the growth of methane and ethane gas hydrates. Chem Eng Sci 49(8):1131–1143
20. Yapa PD, Zheng L, Chen F (2001) A model for deepwater oil/gas blowouts. Mar Pollut Bull 43(7):234–241
21. Wang Z, Yu J, Zhang J et al (2018) Improved thermal model considering hydrate formation and deposition in gas-dominated systems with free water. Fuel 236:870–879
22. Wang ZY, Zhao Y, Sun BJ et al (2016) Heat transfer model for annular-mist flow and its application in hydrate formation risk analysis during deepwater gas-well testing. Chin J Hydrodyn 31(1):20–27
23. Wang Z, Yang Z, Zhang J et al (2018) Quantitatively assessing hydrate-blockage development during deepwater-gas-well testing. SPE Journal 23(4):1166–1183
24. Zhang J, Wang Z, Sun B et al (2019) An integrated prediction model of hydrate blockage formation in deep-water gas wells. Int J Heat Mass Transf 140:187–202
25. Pan L, Hanratty TJ (2002) Correlation of entrainment for annular-mist flow in horizontal pipes. Int J Multiph Flow 28(3):385–408
26. Andreussi P, Asali JC, Hanratty TJ (1985) Initiation of roll waves in gas-liquid flows. AIChE J 31(1):119–126
27. Lorenzo MD, Aman ZM, Kozielski K et al (2014) Underinhibited hydrate formation and transport investigated using a single-pass gas-dominant flowloop. Energy Fuels 28(11):7274–7284
28. Mantilla I, Kouba G, Viana F et al (2012) Experimental investigation of liquid entrainment in gas at high pressure. In: 8th North American conference on multiphase technology. BHR Group, pp 211–225
29. Aman ZM, di Lorenzo M, Kozielski K et al (2016) Hydrate formation and deposition in a gas-dominant flowloop: initial studies of the effect of velocity and subcooling. J Nat Gas Sci Eng 35:1490–1498
30. Wang Z, Zhao Y, Sun B et al (2016) Modeling of hydrate blockage in gas-dominated systems. Energy Fuels 30(6):4653–4666
31. Turner D, Boxall J, Yang S et al (2005) Development of a hydrate kinetic model and its incorporation into the OLGA2000® transient multiphase flow simulator. In: 5th International conference on gas hydrates, Trondheim, Norway, pp 12–16

32. Wang Z, Zhang J, Sun B et al (2017) A new hydrate deposition prediction model for gas-dominated systems with free water. Chem Eng Sci 163(Complete):145–154
33. Wang Z, Zhang J, Chen L et al (2017) Modeling of hydrate layer growth in horizontal gas-dominated pipelines with free water. J Nat Gas Sci Eng 50:364–373
34. Zhang J, Wang Z, Liu S et al (2019) Prediction of hydrate deposition in pipelines to improve gas transportation efficiency and safety. Appl Energy 253:113521
35. Jacob M (1936) Heat transfer in evaporation and condensation II. Mech Eng 58:729–740
36. Majumdar A, Mezic I (1999) Instability of ultra-thin water films and the mechanism of drop formation on hydrophilic surfaces. J Heat Transf 121(4):964–971
37. Haraguchi T, Shimada R, Takeyama T (1989) Drop formation mechanism in dropwise condensation on the polyvinylidene chloride surface. (Proposing a film growth hypothesis). Nihon Kikai Gakkai Ronbunshu B Hen/Trans Jpn Soc Mech Eng Part B 55(519):3472–3478
38. Tammann G, Boehme W (1935) Die Zahl der Wassertröpfchen bei der Kondensation auf verschiedenen festen Stoffen. Ann Phys 414(1):77–80
39. Gorbunov B, Hamilton R (1996) Water nucleation on aerosol particles containing surface-active agents. J Aerosol Sci 27(Suppl 1):S385–S386
40. Bergh S (1998) Water nucleation on aerosol particles containing both organic and soluble inorganic substances. Atmos Res 47(2):271–283
41. Rao I, Koh CA, Sloan ED et al (2013) Gas hydrate deposition on a cold surface in water-saturated gas systems. Ind Eng Chem Res 52(18):6262–6269
42. Zerpa LE, Rao I, Aman ZM et al (2013) Multiphase flow modeling of gas hydrates with a simple hydrodynamic slug flow model. Chem Eng Sci 99(32):298–304
43. Rehder G, Kirby SH, Durham WB et al (2004) Dissolution rates of pure methane hydrate and carbon-dioxide hydrate in undersaturated seawater at 1000-m depth. Geochim Cosmochim Acta 68(2):285–292
44. Ullerich JW, Selim MS, Sloan ED (1987) Theory and measurement of hydrate dissociation. AIChE J 33(5):747–752
45. Selim MS, Sloan ED (2010) Heat and mass transfer during dissociation of hydrates in porous media. AIChE J 35(6):1049–1052
46. Kamath VA, Holder GD, Angert PF (1984) Three phase interfacial heat transfer during the dissociation of propane hydrates. Chem Eng Sci 39(10):1435–1442
47. Kamath VA, Holder GD (1987) Dissociation heat transfer characteristics of methane hydrates. AIChE J 33(2):347–350
48. Kim HC, Bishnoi PR, Heidemann RA et al (1987) Kinetics of methane hydrate decomposition. Chem Eng Sci 42(7):1645–1653
49. Clarke M, Bishnoi PR (2000) Determination of the intrinsic rate of ethane gas hydrate decomposition. Chem Eng Sci 55(21):4869–4883
50. Clarke, MA, Bishnoi R (2001) Measuring and modeling the rate of decomposition of gas hydrates formed from mixtures of methane and ethane. Chem Eng Sci 56(16):4715–4724
51. Sun C, Huang Q, Chen G (2006) Study progress for thermodynamics and dynamics of gas hydrate formation. J Chem Eng 57(5):1031–1039
52. Sun C (2001) Dynamics for hydrate formation/decomposition and related studies. China University of Petroleum, Beijing
53. Makogon TY, Mehta AP, Sloan ED Jr (1996) Structure H and structure I hydrate equilibrium data for 2, 2-dimethylbutane with methane and xenon. J Chem Eng Data 41(2):315–318
54. Jamaluddin AK, Kalogerakis N, Bishnoi PR (1989) Modeling of decomposition of a synthetic core of methane gas hydrate by coupling intrinsic kinetics with heat transfer. Phys Chem 67:945–948
55. Handa YP (1988) A calorimetric study of naturally occurring gas hydrates. Ind Eng Chem Res 27(5):872–874
56. Takeya S, Shimada W, Kamata Y et al (2001) In situ X-ray diffraction measurements of the self-preservation effect of CH_4 hydrate. J Phys Chem A 105(42):9756–9759
57. Takeya S, Ebinuma T, Uchida T et al (2002) Self-preservation effect and dissociation rates of CH_4 hydrate. J Cryst Growth 237(1):379–382

58. Takeya S, Uchida T, Nagao J et al (2005) Particle size effect of hydrate for self-preservation. Chem Eng Sci 60(5):1383–1387
59. Yan L, Liu J (2002) Gas storage capacity of methane hydrate in activated carbon. Acta Petrolei Sin (Pet Process Sect) 18(2):1–5
60. Lin W (2005) Basic studies related to gas mixture separation with hydrate method. China University of Petroleum, Beijing
61. Sun X, Wang Z, Sun B et al (2018) Modeling of dynamic hydrate shell growth on bubble surface considering multiple factor interactions. Chem Eng J 331:221–233
62. Sun X, Sun B, Wang Z et al (2018) A hydrate shell growth model in bubble flow of water-dominated system considering intrinsic kinetics, mass and heat transfer mechanisms. Int J Heat Mass Transf 117:940–950
63. Sun X, Sun B, Wang Z et al (2017) A new model for hydrodynamics and mass transfer of hydrated bubble rising in deep water. Chem Eng Sci 173:168–178

Chapter 3
Prediction for NGH Formation Area in Deepwater Gas Well

The low-temperature and high-pressure environment in deepwater is conducive to the hydrate formation, and the high risks of hydrate deposition and blockage in deepwater gas well/pipelines would do much harm to the on-site operations [1–3]. Therefore, the accurate prediction for NGH formation area in deepwater gas well/pipeline plays a role in analyzing the risks of hydrate deposition and blockage, designing the hydrate management scheme and ensuring the safety of the on-site operations. Currently, based on the relation between temperature and pressure fields distribution in well/pipeline and hydrate phase equilibrium curve, we could make an accurate prediction for the hydrate formation area [4–8].

3.1 Coupling Between Hydrate Behavior and Multi-phase Flow Characteristics

The hydrate formation and deposition have significant impacts on the multi-phase flow and heat transfer in well/pipeline. After hydrate formation, hydrate deposition on the pipewall would reduce the cross-sectional area of the pipe and the flow rate of fluid, surface roughness on pipewall and friction pressure drop have evolved as a result. Moreover, the heat release caused by hydrate formation would have an influence on the heat transfer in the system [7, 9–12]. Conversely, the free water and its content in the system, flow pattern and contact relationship of gas-liquid phase all have a significant impact on the behavior of hydrate formation and deposition. Besides, under a high flow rate of fluid, hydrate deposition is less likely to be observed. Therefore, the coupling relations between hydrate behavior (formation, migration and deposition of hydrate particles) and the characteristics of fluid dynamics and heat transfer in multiphase flow, shown in Fig. 3.1, should be taken into account for establishing the multiphase flow model.

The physical model of flow and heat transfer in well during well test is shown in Fig. 3.2, where the whole system is composed of the wellbore, casing pipe, riser,

© Science Press and Springer Nature Singapore Pte Ltd. 2020
Z. Wang et al., *Natural Gas Hydrate Management in Deepwater Gas Well*,
https://doi.org/10.1007/978-981-15-6418-5_3

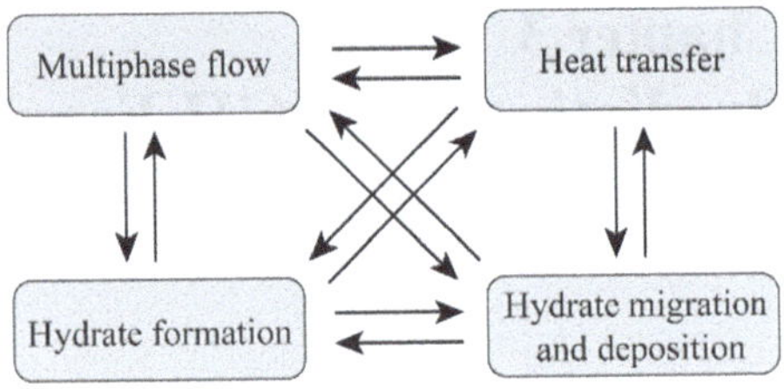

Fig. 3.1 Coupling relation in multiphase flow

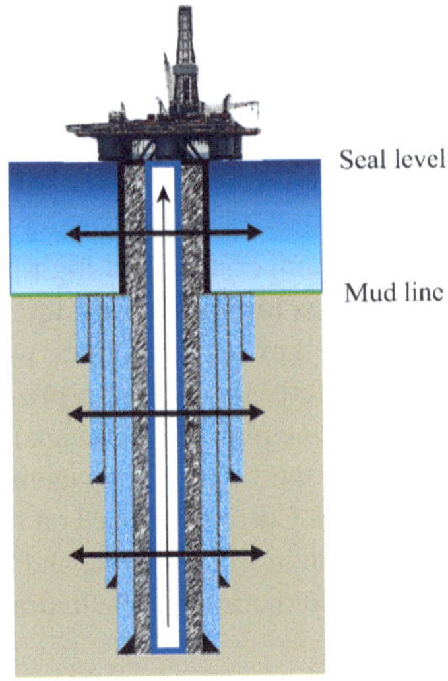

Fig. 3.2 Physical model of flow and heat transfer in well

well-casing well annulus (filled with well fluid), casing-casing well annulus (filled with cement or drilling fluid), well-riser well annulus (filled with drilling-completion fluid), casing-stratum well annulus (filled with cement), stratum and seawater.

During production and test, the great pressure variation and strong gas compressibility in wellbore/pipeline would make the gas-liquid ratio and liquid film thickness vary with well depth. What's more, the flow pattern in gas well usually present as the annular-mist flow, in which the diagram of fluid flow and heat transfer below mudline is shown in Fig. 3.3. The expressive modes of the liquid phase in well/pipeline mainly include the liquid film on pipewall and droplets in the gas core. More concretely, the instability of disturbance wave on the gas-liquid interface, the properties difference and the lowest energy principle between the gas and liquid phase would promote the separation of gas and liquid phase. By now, the gas phase with high velocity flows along the center of the pipe forming the continuous gas core; the liquid phase with low velocity transforms a layer of annular liquid film on pipewall, while some droplets in the liquid film could enter the gas core under shearing action and flow with the gas phase. Therefore, there are two liquid phase modes of liquid film and droplets. Besides, the droplets just account for a large proportion, and its flow velocity

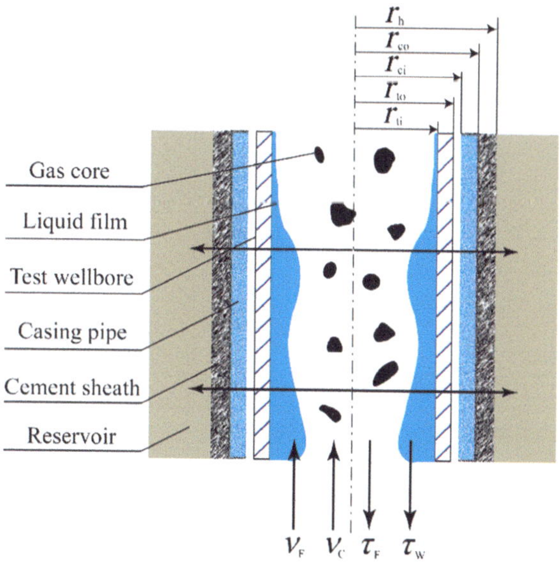

Fig. 3.3 Diagram of fluid flow and heat transfer below mudline in annular-mist flow

is approximately equal to that of the gas phase, so the droplets tend to make similar characteristics of flow and heat transfer with the gas phase [13].

3.2 Prediction on Well/Pipeline Temperature and Pressure Fields in Annular-Mist Flow

3.2.1 Multiphase Flow Model

3.2.1.1 Model Assumption

To accurately calculate the temperature and pressure fields of a wellbore, the multiphase flow model of well should take into account the hydrate formation and deposition, which includes the continuity equation, momentum equation and energy equation [14]. The process of fluid flow and heat transfer in well is shown in Fig. 3.4 [11]. This model is assumed:

(1) The proportion of dissolved gas to the total gas is small;
(2) Hydrate is located in the hydrate phase equilibrium region after hydrate formation;
(3) The temperature and pressure remain constant when the formation fluid enters into well;
(4) The fluid in well belongs to one-dimensional steady flow;

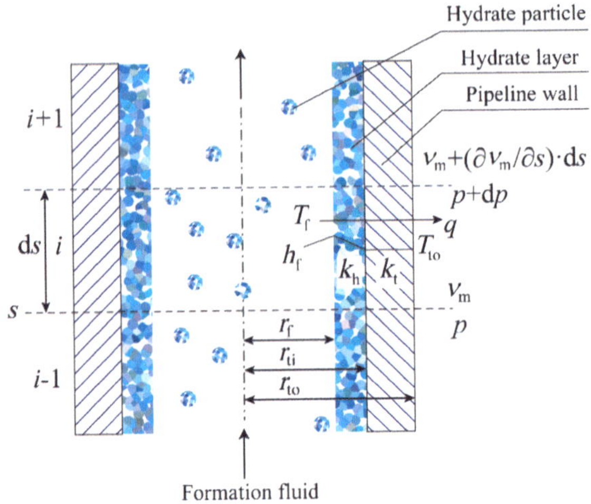

Fig. 3.4 Diagram of flow and heat transfer considering hydrate formation and deposition

(5) The heat transfer of fluid in well is stable, but the heat transfer of fluid in formation and seawater is unstable;
(6) The distribution of seawater temperature is known, and the formation temperature is a linear function of well depth;
(7) The well and casing are concentric;
(8) There is no gas phase in the liquid film on pipewall, and the axial pressure gradient in the gas core is the same as that in the liquid film.

3.2.1.2 Continuity Equation

Based on the above diagram, set the upward direction of well as the positive direction of coordinate and consider the well bottom as the coordinate origin. The flow pattern in the well would transform from the gas-liquid two phases flow to the gas-liquid-solid three phases flow due to the hydrate formation. Furthermore, the gas compressibility and process of hydrate formation and deposition would make the flow transient. In microelement, the continuity equations of gas, liquid and hydrate phases are shown in Eqs. 3.1, 3.2, and 3.3 [15].

$$\frac{\partial}{\partial t}(\rho_g A E_g) + \frac{\partial}{\partial s}(\rho_g A v_g E_g) = -x_g r_{hf} \tag{3.1}$$

$$\frac{\partial}{\partial t}(A\rho_l E_l) + \frac{\partial}{\partial s}(A\rho_l v_l E_l) = -(1 - x_g) r_{hf} \tag{3.2}$$

$$\frac{\partial}{\partial t}(A\rho_h E_h) + \frac{\partial}{\partial s}(A\rho_h v_h E_h) = r_{hf} - r_{hd} \tag{3.3}$$

$$x_g = \frac{M_g}{M_g + N \cdot M_{H_2O}} \tag{3.4}$$

where r_{hf} refers to the hydrate formation rate per length of pipeline, kg/(s·m); r_{hd} refers to the hydrate deposition rate per length of pipeline, kg/(s·m); E_g, E_l, E_h show the volume fraction of gas, liquid, and hydrate phase (not including the part deposited and attached on pipewall), %; V_g, V_l, V_h show the movement velocity of gas, fluid, and hydrate phase, m/s; A shows the effective section area of pipeline, m^2; x_g is the mass fracture of gas in hydrate, calculated by Eq. 3.4; N represents the hydration number; taking the mean value of 6; ρ_l, ρ_h show the density of fluid and hydrate phase, kg/cm^3; ρ_g stands for the gas density, slightly affected by temperature and pressure, kg/cm^3, considered as a constant.

Hydrate deposition would reduce the effective section area of pipeline, which is calculated as shown in Eq. 3.5.

$$A = \pi r_f^2 \tag{3.5}$$

$$r_f = r_{ti} - \delta_D \tag{3.6}$$

where r_f refers to the effective inner diameter of pipeline, calculated by Eq. 3.6, m; r_{ti} is the initial inner diameter of pipeline, m; δ_D shows the hydrate layer thickness on pipewall, m.

3.2.1.3 Momentum Equation

Fluid pressure in a static state is equal to the difference between stratum pressure and hydrostatic pressure at one depth. Frictional resistance, gravity, and acceleration need to be overcome while fluid flows, which could reduce the fluid pressure. The distribution of fluid pressure could be solved by the momentum equation.

Suppose that the gravity and the slippage loss of gas-liquid two-phase flow are neglected, the vertical and inclined flow during production are considered, momentum equation is established as shown in Eq. 3.7 [16]. Its integration could obtain the pressure drop in well, which represents the impact of hydrate deposition on pressure drop.

$$-\frac{dp}{ds} = \frac{\partial}{\partial t}(A\rho_m v_m) + \frac{\partial}{\partial s}(A\rho_m v_m^2) + Ag\rho_m \cos\theta + \left| f_F \frac{\rho_m v_m^2}{2d_f} \right| \tag{3.7}$$

$$\rho_m = \rho_g E_g + \rho_l E_l + \rho_h E_h \tag{3.8}$$

$$f_F = \left\{ -1.737 \ln\left[0.269\frac{\varepsilon}{D} - \frac{2.185}{Re} \ln\left(0.269\frac{\varepsilon}{D} + \frac{14.5}{Re} \right) \right]^{-2} \right\} \tag{3.9}$$

where p is the pipeline pressure, MPa; θ is the deviation angle of the wellbore,(°); ρ_m shows the density of mixture including gas, liquid, and hydrate particle, calculated by Eq. 3.8; v_m shows the movement velocity of the mixture, m/s; f_F stands for the friction coefficient, a function of surface roughness and Reynolds number, calculated by Eq. 3.9; ε represents the surface roughness of well, approximately equal to the value of δ_D.

3.2.1.4 Energy Equation

(1) Calculate the Temperature Field of Seawater

The seawater temperature is indispensable for the model solution. Seawater could be divided into the mixed layer, thermocline and constant temperature layer in a vertical direction. Specifically, the mixed layer generally exists within 100 m of the ocean surface, which has uniform water temperature and small vertical temperature gradient; the thermocline layer is located below mixed layer and above constant temperature layer with large vertical temperature gradient; the constant temperature layer, with water temperature ranging from 2 °C to 6 °C, is located below thermocline layer until the sea bottom. Gao Yonghai et al. [17] obtained the function relationship of seawater temperature and water depth by fitting, which is shown in Eq. 3.10 when the water depth is greater than 200 m. However, the seawater temperature distribution depends on the season when the water depth is less than 200 m. Their function relationship in spring, summer, autumn and winter is shown in Eqs. 3.11–3.17.

$$T_{sea} = a_1 + a_2 \Big/ \left[1 + e^{(h+a_0)/a_3}\right], h > 200 \text{ m} \tag{3.10}$$

Spring:

$$T_{sea} = \frac{T_S(200 - h) + 13.68h}{200}, 0 \le h < 200 \text{ m} \tag{3.11}$$

Summer:

$$T_{sea} = T_S, 0 \le h < 20 \text{ m} \tag{3.12}$$

$$T_{sea} = \frac{T_S(200 - h) + 13.7(h - 20)}{180}, 20 \text{ m} \le h < 200 \text{ m} \tag{3.13}$$

Autumn:

$$T_{sea} = T_S, 0 \le h < 50 \text{ m} \tag{3.14}$$

$$T_{sea} = \frac{T_S(200 - h) + 13.7(h - 50)}{150}, 50 \text{ m} \le h < 200 \text{ m} \tag{3.15}$$

Winter:

$$T_{sea} = T_S, 0 \le h < 100\ \text{m} \tag{3.16}$$

$$T_{sea} = \frac{T_S(200 - h) + 13.7(h - 100)}{100}, 100\ \text{m} \le h < 200\ \text{m} \tag{3.17}$$

where $a_1 = 3.940, a_2 = 37.091, a_0 = 130.137, a_3 = 402.732$; T_{sea} shows the seawater temperature, °C; T_S shows the temperature on the seawater surface, °C; h refers to the seawater depth, m.

(2) Calculate the Heat Loss Per Well Length in Section of Stratum and Seawater

Based on the temperature difference above and below mudline, the deepwater wellbore could be divided into the upper part (seawater section) and the lower part (stratum section). There are three heat transfer methods in deepwater well including heat conduction, heat convection, and heat radiation. The heat transfer process is shown in Fig. 3.5.

Determining the temperature profile of wellbore is the key to calculate the hydrate formation rate and predict the dynamics of hydrate blockage. Considering the coupling in the multiphase flow system, based on the energy conservation, the energy equation of well fluid is established in Eq. 3.18.

$$W_m\left(\frac{dH_f}{ds} + v_j\frac{dv}{ds} - g\sin\theta\right) - \frac{R_{hf}\Delta H}{M_h} = -q \tag{3.18}$$

$$\frac{dH_f}{ds} = C_m\frac{dT_f}{ds} - \mu_j C_m\frac{dp}{ds} \tag{3.19}$$

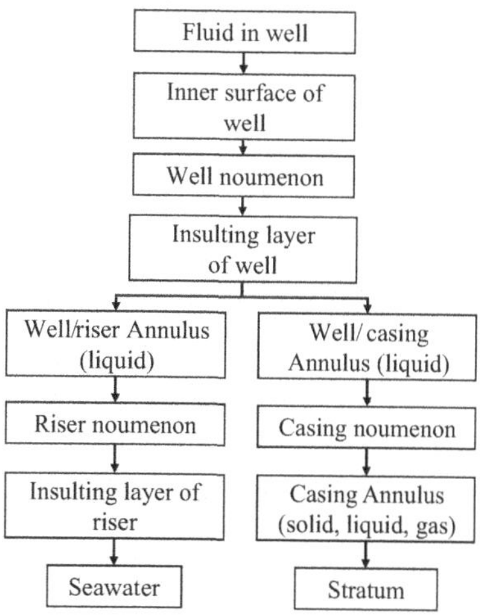

Fig. 3.5 Heat transfer process in deepwater gas well

$$C_m = \frac{q_g \rho_g}{q_g + q_l} C_g + \frac{q_l \rho_l}{q_g + q_l} C_l \tag{3.20}$$

$$\mu_j = \left(\frac{\partial T_f}{\partial p}\right)_H = \frac{1}{C_g}\left[T_f\left(\frac{\partial V}{\partial T_f}\right)_p - V\right] = -\frac{1}{C_g \rho_g}\frac{T_f}{Z_g}\left(\frac{\partial Z_g}{\partial T_f}\right)_p \tag{3.21}$$

$$q = \frac{1}{A'}(T_f - T_{to}) \tag{3.22}$$

$$A' = \frac{1}{2\pi r_{to} U_{to}} \tag{3.23}$$

$$U_{to} = \frac{r_{to}}{r_{ti} h_f} + \frac{r_{to} \ln\left(\frac{r_{ti}}{r_f}\right)}{k_h} + \frac{r_{to} \ln\left(\frac{r_{to}}{r_{ti}}\right)}{k_t} \tag{3.24}$$

where W_m shows the mass flow of fluid, kg/s; H_f refers to the specific enthalpy of mixture; the specific enthalpy gradient could be expressed by temperature and pressure gradient shown in Eq. 3.19; θ is the deviation angle of wellbore, (°); r_{hf} represents the hydrate formation rate per length of pipeline, kg/(s·m); ΔH shows the enthalpy of hydrate formation, J/mol; M_h is the mean molar mass of hydrate, kg/mol; C_m is the specific heat of mixture, J/(kg·K); calculated by Eq. 3.20 by neglecting the heat capacity of hydrate; μ_j is the Joule–Thomson coefficient of test fluid, calculated by Eq. 3.21; T_f is the fluid temperature in well, °C; q represents the lost heat per well length, J, calculated by Eq. 3.22; T_{to} shows the outer surface temperature of pipeline, °C; $\frac{1}{A'}$ stands for the relaxation parameter, calculated by Eq. 3.23; r_f is the effective inner diameter of pipeline, m; r_{ti} is the initial inner diameter of pipeline, m; r_{to} is the outer diameter of pipeline, m; U_{to} stands for the comprehensive heat transfer coefficient of pipeline, W/(m^2·K), calculated by Eq. 3.24; k_h, k_t are the respective thermal conductivity of hydrate and pipeline, W/(m·K).

The heat exchange processes of well in stratum section include forced-convection heat transfer between fluid and hydrate layer surface, heat conduction of hydrate layer, heat conduction of well wall, well annulus radiation heat transfer, natural convection heat transfer, heat conduction of casing wall, heat conduction of cement ring and the unsteady heat conduction in the stratum. The heat transfer equation between well fluid and stratum is shown in Eq. 3.25.

$$q = \frac{1}{A'}(T_f - T_{ei}) \tag{3.25}$$

$$\frac{1}{A'} = \frac{2\pi r_{to} U_{to} k_e}{k_e + r_{to} U_{to} T_D} \tag{3.26}$$

$$\frac{1}{U_{\mathrm{to}}}=\frac{r_{\mathrm{to}}}{r_{\mathrm{f}}h_{\mathrm{i}}}+\frac{r_{\mathrm{to}}\ln\left(r_{\mathrm{ti}}/r_{\mathrm{f}}\right)}{k_{\mathrm{h}}}+\frac{r_{\mathrm{to}}\ln\left(r_{\mathrm{to}}/r_{\mathrm{ti}}\right)}{k_{\mathrm{t}}}+\frac{1}{h_{\mathrm{c}}+h_{\mathrm{r}}}+\frac{r_{\mathrm{to}}\ln\left(r_{\mathrm{co}}/r_{\mathrm{ci}}\right)}{k_{\mathrm{c}}}+\frac{r_{\mathrm{to}}\ln\left(r_{\mathrm{wb}}/r_{\mathrm{co}}\right)}{k_{\mathrm{cem}}} \tag{3.27}$$

where $\frac{1}{A'}$ shows the relaxation parameter, calculated by Eq. 3.26; T_{f} is the fluid temperature in well, °C; T_{ei} is the stratum/seawater temperature, °C; T_{D} is the dimensionless temperature; k_{e} refers to the thermal conductivity of stratum, W/(m·K); U_{to} represents the comprehensive heat transfer coefficient of wellbore, W/(m^2·K), calculated by Eq. 3.27; r_{f} is the effective inner diameter of pipeline, m; r_{ti} is the initial inner diameter of pipeline, m; r_{to} is the outer diameter of pipeline, m; r_{wb} is the borehole radius, m; r_{ci} is the inner diameter of casing, m; r_{co} is the outer diameter of the casing, m; h_{i} refers to the convective heat transfer coefficient, W/(m^2·K); h_{c} refers to the heat transfer coefficient of seawater, W/(m^2·K); k_{h}, k_{c}, k_{cem} show the thermal conductivity of hydrate, casing, and cement casing, W/(m·K).

Substituting Eqs. 3.25–3.27 into Eq. 3.18 could obtain the temperature gradient equation of well fluid in stratum section as below:

$$\frac{\mathrm{d}T_{\mathrm{f}}}{\mathrm{d}s}=\frac{1}{w_{\mathrm{f}}C_{\mathrm{m}}}\left[\frac{1}{A'}(T_{\mathrm{f}}-T_{\mathrm{ei}})+\frac{R_{\mathrm{hf}}\Delta H}{M_{\mathrm{h}}}\right]+\mu_{\mathrm{j}}\frac{\mathrm{d}p}{\mathrm{d}s}-\frac{1}{C_{\mathrm{m}}}\left(-g\sin\theta+v\frac{\mathrm{d}v}{\mathrm{d}s}\right) \tag{3.28}$$

The heat exchange process of well in seawater section includes forced-convection heat transfer between well fluid and hydrate layer surface, hydrate layer heat conduction, well wall heat conduction, heat transfer by well annulus radiation, heat transfer by natural convection, riser wall heat conduction and convective heat transfer between outer surface of riser and seawater. The heat transfer equation between well fluid and seawater is shown in Eq. 3.29.

$$q=\frac{1}{B'}(T_{\mathrm{f}}-T_{\mathrm{sea}}) \tag{3.29}$$

$$\frac{1}{B'}=2\pi r_{\mathrm{to}}U_{\mathrm{to}} \tag{3.30}$$

$$\frac{1}{U_{\mathrm{to}}}=\frac{r_{\mathrm{to}}}{r_{\mathrm{f}}h_{\mathrm{i}}}+\frac{r_{\mathrm{to}}\ln\left(r_{\mathrm{ti}}/r_{\mathrm{f}}\right)}{k_{\mathrm{h}}}+\frac{r_{\mathrm{to}}\ln\left(r_{\mathrm{to}}/r_{\mathrm{ti}}\right)}{k_{\mathrm{t}}}+\frac{1}{h_{\mathrm{c}}+h_{\mathrm{r}}}+\frac{r_{\mathrm{to}}\ln\left(r_{\mathrm{co}}/r_{\mathrm{ci}}\right)}{k_{\mathrm{c}}}+\frac{r_{\mathrm{to}}}{r_{\mathrm{co}}h_{\mathrm{s}}} \tag{3.31}$$

where $\frac{1}{B'}$ shows the relaxation parameter, calculated by Eq. 3.30; T_{f} is the fluid temperature in well, °C; T_{sea} is the seawater temperature, °C; U_{to} refers to the comprehensive heat transfer coefficient of well, W/(m^2·K), calculated by Eq. 3.31; r_{co} represents the outer diameter of the casing, m; r_{ci} represents the inner diameter of the casing, m; r_{to} represents the outer diameter of pipeline, m; r_{ti} represents the initial inner diameter of pipeline, m; h_i stands for the convective heat transfer coefficient, W/(m^2·K); h_c stands for the heat transfer coefficient of seawater, W/(m^2·K); k_c, k_h refer to the thermal conductivity of casing and hydrate, W/(m·K); h_s refers to the forced-convection heat coefficient between outer surface of riser and seawater, W/(m^2·K).

Substituting Eqs. 3.29–3.31 into Eq. 3.18 could obtain the temperature gradient of well fluid in seawater section as below:

$$\frac{dT_f}{ds} = \frac{1}{w_f C_m}\left[\frac{1}{B'}(T_f - T_{sea}) + \frac{R_{hf}\Delta H}{M_h}\right] + \mu_j \frac{dp}{ds} - \frac{1}{C_m}\left(-g\sin\theta + v\frac{dv}{ds}\right) \tag{3.32}$$

Under the working condition of a well shut-in, well fluid takes heat from the stratum section or dissipates heat in the seawater section, which directly determines the temperature variation of drilling fluid. Based on the impact of hydrate formation and deposition in well/pipeline on heat transfer, the energy balance relation per unit length of drilling fluid column could be expressed as Eq. 3.33, which considers the phase transition heat caused by hydrate formation, thermal resistance variation caused by hydrate deposition and Joule-Thomson effect caused by the change of pipe diameter.

$$\rho c_p \frac{\partial T}{\partial t}\pi r_{ti}^2 = \frac{k(T_e - T)\cdot 2\pi r_t}{r_{to} - r_{ti}} \tag{3.33}$$

3.2.2 Model Solution

The finite difference method tends to be used to solve the multiphase flow model due to its unsteady characteristics. The solution steps are shown in Fig. 3.6 [11]. Notice that the appropriate conditions of the definite solution shall be established under different working conditions.

The specific solution steps are as follows:

(1) Divide the well/pipeline into multiple sections to obtain several elements; Suppose that the hydrate formation and deposition rate, δ_D and surface roughness of pipe per element remains unchanged.
(2) The inlet temperature and pressure, as well as each-phase volume fraction of element i at time n, are known;
(3) Assume the outlet temperature T_{out} and pressure p_{out} of element i at time n;
(4) Take the mean value of temperature and pressure $p_{mean} = (p_{in} + p_{out})/2$, $T_{mean} = (T_{f,in} + T_{f,out})/2$;
(5) Calculate the gas density, hydrate formation and deposition rate, δ_D, the effective section area of pipeline and frictional pressure drop of pipeline according to the above p_{mean} and T_{mean}.

Fig. 3.6 Flowsheet of model solution

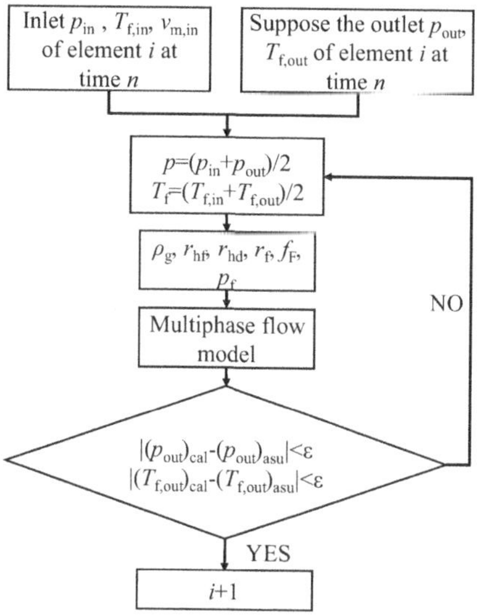

Well/pipeline blockage would be thought to happen if the effective radius of the pipe r_f is smaller than the critical value r_c;

(6) Substitute the above parameters into the multiphase flow model to obtain the outlet temperature and pressure of element i and compare them with the assumed temperature and pressure.
The calculation would finish if the error requirement is satisfied. Otherwise, repeat the above calculation process;
(7) Consider the outlet pressure value of element i at time n as the inlet parameter of element $i + 1$ at time n, then perform the same calculation process to obtain parameters of element $i + 1$ such as δ and the effective diameter of the pipe;
(8) Use the temperature and pressure values per element at time n as the initial values, then perform the same calculation to obtain the parameters at time $n + 1$.

Through the solution above, the temperature and pressure distribution of fluid, δ and the effective diameter of the pipe could be obtained.

3.2.3 Model Validation

With the multiphase flow model, the temperature and pressure field of wellbores #1 and #2 in one sea area could be calculated and compared with the results calculated by others to verify the reliability of the model. The basic parameters of well #1 and well #2 are shown in Table 3.1 [18] and Table 3.2 [19] respectively.

Table 3.1 Basic parameters of well #1 in one sea area

Well depth/m	Depth of the sea/m	Bottom hole pressure/MPa	Bottom hole temperature/°C	Seabed temperature/°C	Sea level temperature/°C	Inner diameter of well/mm
2896.6	763	20.6	82	4.4	15.6	76
Comprehensive heat transfer coefficient of well below seabed/W $(m^2 \cdot k)^{-1}$	Comprehensive heat transfer coefficient of well above seabed/W $(m^2 \cdot k)^{-1}$	Heat transfer coefficient of stratum/W $(m \cdot k)^{-1}$	Heat dissipation coefficient of stratum/$10^{-6} m^2 \cdot s^{-1}$	Daily gas output/10^4 $m^3 \cdot d^{-1}$	Volume water content/‰	
5.678	11.356	2.2	1	27	1	

Table 3.2 Parameters of well #2 in one sea area

Well depth/m	Seabed depth/m	Bottom hole pressure/MPa	Bottom hole temperature/°C	Seabed temperature/°C	Sea level temperature/°C
4450	853.4	28.84	73.33	1.11	12.22
Comprehensive heat transfer coefficient of well below seabed/W $(m^2 \cdot k)^{-1}$	Comprehensive heat transfer coefficient of well above seabed/W $(m^2 \cdot k)^{-1}$	Inner diameter of wellbore/mm	Daily liquid output/$m^3 \cdot d^{-1}$	Air content/$Nm^3 \cdot m^3$	Water content/%
5.6785	11.356	101.6	500	195.8	49

The temperature results of wellbore #1 and #2 are shown in Figs. 3.7 and 3.8 respectively, illustrating the results calculated by the multiphase flow model are highly consistent with the results calculated by Alves et al. [18] and Hasan et al. [19], and the maximum deviation is less than 5%.

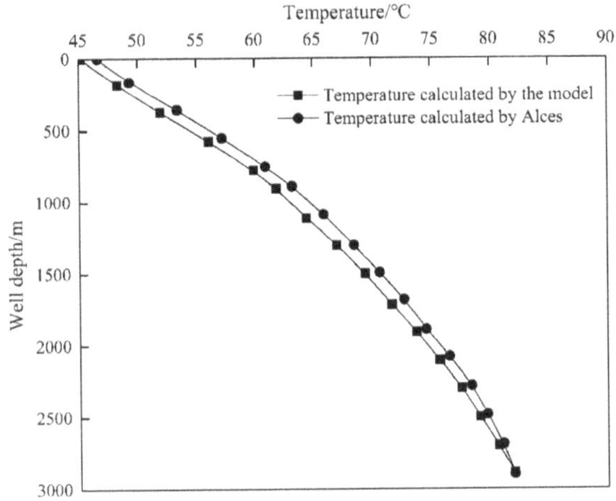

Fig. 3.7 Temperature of wellbore #1

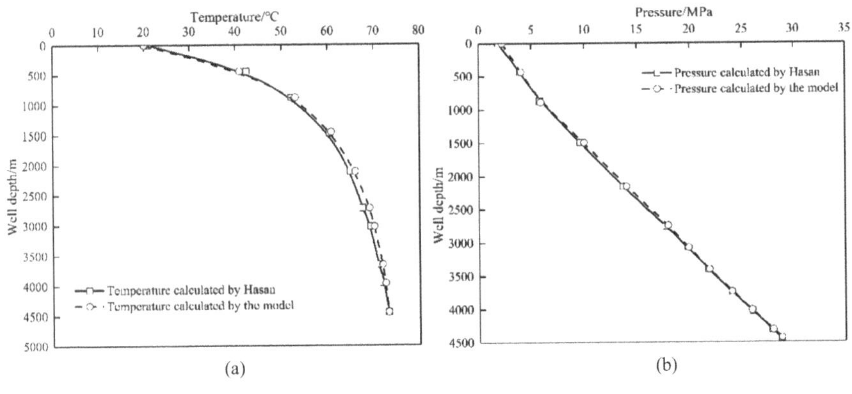

Fig. 3.8 Temperature and pressure of wellbore #2, **a** temperature-depth zone, **b** pressure-depth zone

3.3 Prediction on Well/Pipeline Temperature and Pressure Field in Water-Saturated Gas Systems

Considering the coupling relations between hydrate behavior (formation, migration, and deposition of hydrate particles) and the characteristics of fluid dynamics and heat transfer is favorable to establish the coupling model of temperature and pressure field in deepwater well/pipeline under water-saturated gas single-phase flow. The basic assumptions are as below:

(1) The heat transfer between well fluid and surrounding environment belongs to the one-dimensional radial heat transfer;
(2) The heat transfer of well/pipeline fluid is stable;
(3) Hydrate exists in the hydrate formation area;
(4) The deformation of well/pipeline caused by temperature and pressure variation is neglected.

3.3.1 *Temperature Field Model*

Well/pipeline temperature would decrease along the fluid flow direction in water-saturated gas systems due to the heat exchange between high-temperature fluid and external environment. The temperature distribution in well/pipeline is shown in Fig. 3.9 [20]. Figure 3.9a represents the radial temperature distribution in pipe, where the temperature gradually decreases from the inside to the outside; while Fig. 3.9b stands for the temperature distribution along the axial direction, where the temperature gradually decreases along the direction of fluid flow. Also, there is a thermal boundary layer near the cold wall in the pipe, where the temperature decreases rapidly. Based on the energy balance principle, the energy conservation equation of well/pipeline fluid is established as below [7]:

$$W_{\mathrm{m}}\left(C_{\mathrm{p}}\frac{\mathrm{d}T_{\mathrm{B}}}{\mathrm{d}s} - \mu_{\mathrm{j}}C_{\mathrm{p}}\frac{\mathrm{d}p}{\mathrm{d}s} + U_{\mathrm{g}}\frac{\mathrm{d}U_{\mathrm{g}}}{\mathrm{d}s} - g\sin\theta\right) = -q_1 + q_2 \tag{3.34}$$

where μ_{j} is the Joule–Thomson coefficient; U_{g} shows the flow rate of fluid, m/s; W_{m} shows the mass flow of fluid, kg/s; T_{B} is the fluid temperature, K; q_1 represents the

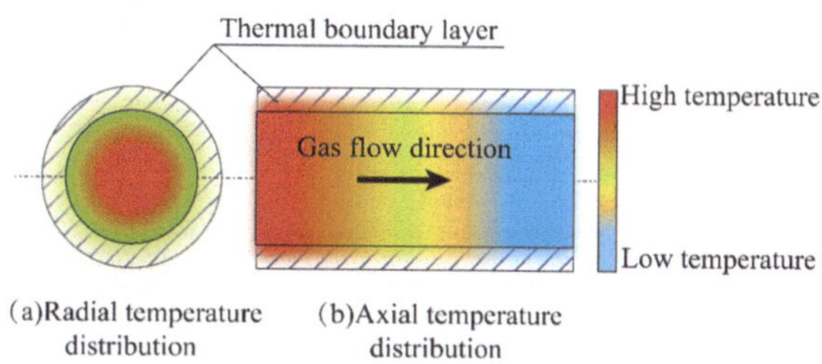

Fig. 3.9 Temperature distribution in well/pipeline under water-saturated gas systems (To develop a better understanding, the thickness of thermal boundary layer is enlarged in the diagram)

heat transfer between fluid and environment, W/m; q_2 stands for the heat released by hydrate formation, W/m; C_p refers to the fluid specific heat in the main flow area of well/pipeline, J/(kg·K).

By neglecting the thermal capacity of hydrate, C_p could be expressed as Eq. 3.35, and the heat loss per length of well/pipeline could be expressed as Eq. 3.36.

$$C_p = \frac{q_g \rho_g}{q_g + q_l} C_g + \frac{q_g \rho_l}{q_g + q_l} C_l \tag{3.35}$$

$$q_l = 2\pi r_c U (T_B - T_e) \tag{3.36}$$

$$\frac{1}{U} = \frac{r_{to}}{r_{ti} h_B} + \frac{r_{to} \ln(r_{ti}/r_i)}{k_h} + \frac{r_{to} \ln(r_{to}/r_{ti})}{k_t} \tag{3.37}$$

$$h_B = \frac{Nu k_B}{D_i} = 0.023 \frac{k_B}{D_i} Re^{4/5} Pr^{1/3} \tag{3.38}$$

where q_l shows the heat loss in dl of pipeline, W/m; T_B is the fluid temperature, K; T_e is the environment temperature outside the pipeline, K; U stands for the comprehensive heat transfer coefficient, W/(m·K); expressed as Eq. 3.37 by taking the outer pipewall as the reference surface; r_{ti} is the initial inner diameter of the pipeline, m; r_{to} is the initial outer diameter of the pipeline, m; k_h represents the heat transfer coefficient of hydrate layer, W/(m·K), which is not sensitive to the temperature and pressure, approximate to constant of 0.5; k_t represents the heat transfer coefficient of the pipeline, W/(m·K); h_B represents the internal heat transfer coefficient, W/(m·K), which changes with pipe diameter and Reynolds number, calculated by Eq. 3.38; k_B refers to the heat transfer coefficient of fluid in main flow area, W/(m·K); Re shows the Reynolds number; Nu shows the Nusselt number; Pr shows the Prandtl number.

The heat released by hydrate formation, which is proportional to the hydrate formation rate and calculated by Eq. 3.39, would make the system temperature rise.

$$q_2 = \frac{\Delta H}{M_h} \cdot \frac{\partial m_l}{\partial t \partial s} \tag{3.39}$$

where ΔH shows the enthalpy of hydrate formation in the pipeline, J/mol; M_h is the molar mass of hydrate, kg/mol; $\frac{\partial m_l}{\partial t \partial s}$ stands for the hydrate formation rate, kg/(s·m).

The thermal resistance of the hydrate layer increases continuously with its growth. Substituting Eqs. 3.36 and 3.39 into the energy conservation Eq. 3.34 could gain the temperature gradient of fluid as below:

$$\frac{dT_B}{ds} = \frac{q_l + q_2}{w_f C_p} + \mu_j \frac{dp}{ds} - \frac{1}{C_p}\left(-g \sin\theta + U_g \frac{dU_g}{ds}\right) \tag{3.40}$$

3.3.2 Pressure Field Model

Taking the inlet of fluid in the deepwater pipeline as the coordinate origin, and considering the flow direction of the fluid in the main flow area in a positive direction, the pipeline coordinate system could be established as shown in Fig. 3.10.

The hydrate layer affects the section area, pipewall roughness and transient flow characteristics of the fluid, and the slippage loss between gas and liquid phases needs to be neglected. Based on the pipeline coordinate and the calculation of pressure drop in horizontal pipe flow, momentum equation per dl could be established shown in Eq. 3.41, where the third item on the left refers to the pressure gradient of fluid.

$$\frac{\partial}{\partial t}\left(A\rho U_{\mathrm{g}}\right) + \frac{\partial}{\partial s}\left(A\rho U_{\mathrm{g}}^{2}\right) + \frac{\mathrm{d}p}{\mathrm{d}s} + \left|32 f_{\mathrm{F}}\frac{W_{\mathrm{m}}^{2}}{\rho\pi^{2}D_{\mathrm{i}}^{5}}\mathrm{d}s\right| = 0 \tag{3.41}$$

$$f_{\mathrm{F}} = \left\{-1.737\ln\left[0.269\frac{\varepsilon}{D_{\mathrm{i}}} - \frac{2.185}{Re\ln\left(0.269\frac{\varepsilon}{D_{\mathrm{i}}} + \frac{14.5}{Re}\right)}\right]^{-2}\right\} \tag{3.42}$$

where f_{F} shows the Fanning friction resistance coefficient, calculated by Eq. 3.42; ε represents the surface roughness.

3.4 Influencing Factors of NGH Phase Equilibrium in Well/Pipeline

The phase equilibrium curve of methane hydrate is shown in Fig. 3.11, illustrating the temperature has a great influence on the critical pressure for hydrate formation. The higher temperature means the higher critical pressure for hydrate formation, and vice versa. Besides, there is one critical temperature for hydrate formation. If the environment temperature is higher than the critical temperature, no hydrate would be found regardless of the pressure. The critical temperatures of various gas hydrates are shown in Table 3.3.

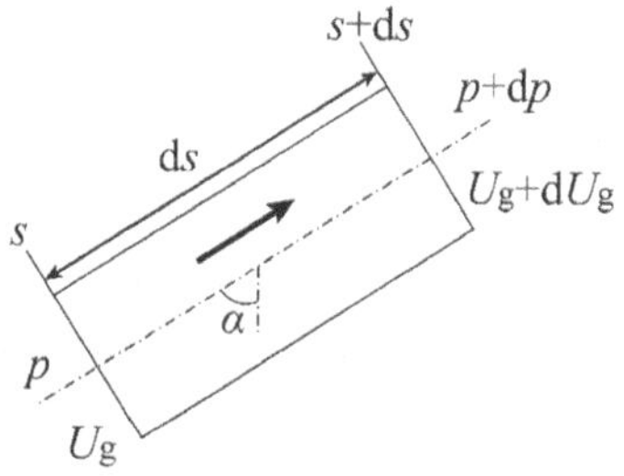

Fig. 3.10 Pipeline coordinate diagram

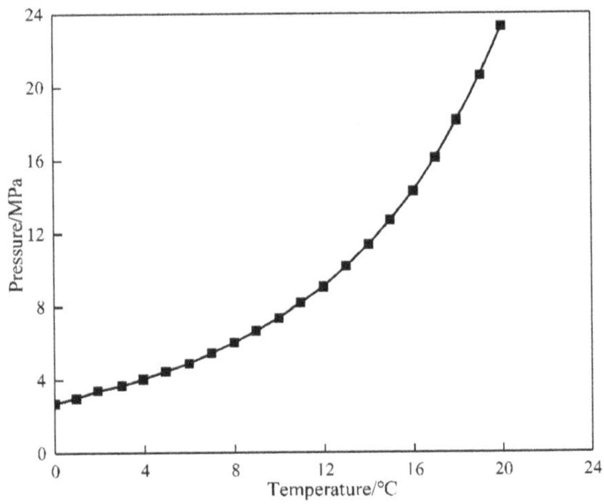

Fig. 3.11 Methane hydrate phase equilibrium curve

Table 3.3 Critical temperature of NGH formation

Gas component	CH_4	C_2H_6	C_3H_8	i-C_4H_{10}	CO_2	H_2S
Critical temperature for hydrate formation/°C	21.5	14.5	5.5	2.5	10.0	29.0

3.4.1 Natural Gas Components

The impact of four natural gas components on hydrate phase equilibrium conditions is analyzed. The temperature and pressure for hydrate formation under different contents of methane, ethane, propane, and *n*-butane are shown in Fig. 3.12, illustrating different gas components have different sensitivity for hydrate formation. With the contents of ethane, propane and *n*-butane increase, all the phase equilibrium curves of hydrate formed by the mixed gas move to the right, indicating such three components are more likely to promote hydrate formation than methane. In addition, propane has a larger promoting effect for hydrate formation than ethane, and *n*-butane and propane have the same promoting effect. For large holes of type II hydrate, the heavy hydrocarbon component of *n*-butane has better stabilizing effects than the light hydrocarbon molecules, so its lower content still could form the type II of the hydrate.

3.4.2 Natural Gas Density

The laboratory experiments for hydrate formation under different gas components require a heavy workload. Therefore, to reduce the amount of experimental work, the impact of natural gas components on hydrate formation tends to be converted into the impact of the density ratio of natural gas. The hydrate phase equilibrium curves

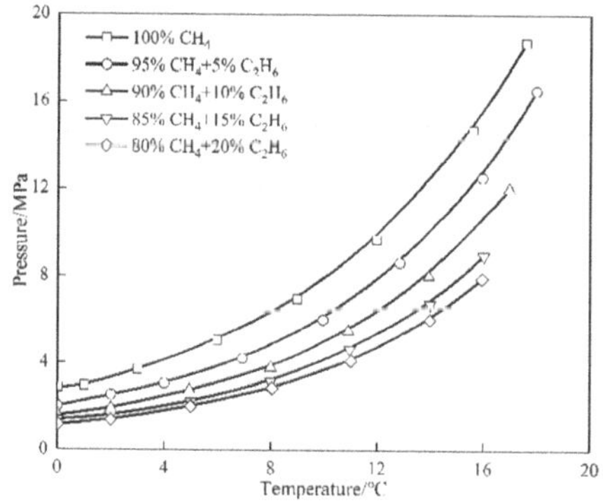

(a) Different contents of Ethane

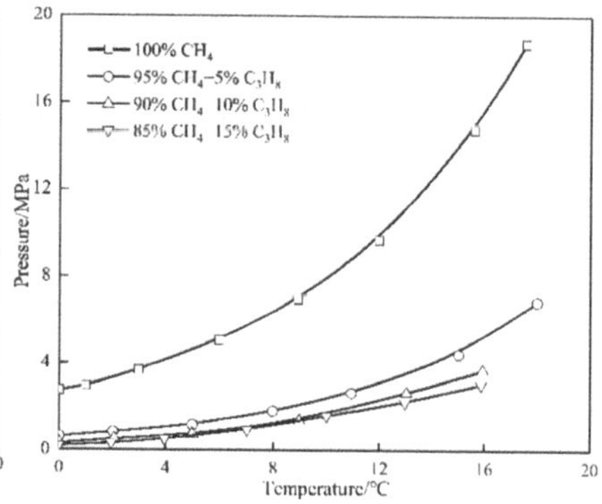

(b) Different contents of propane

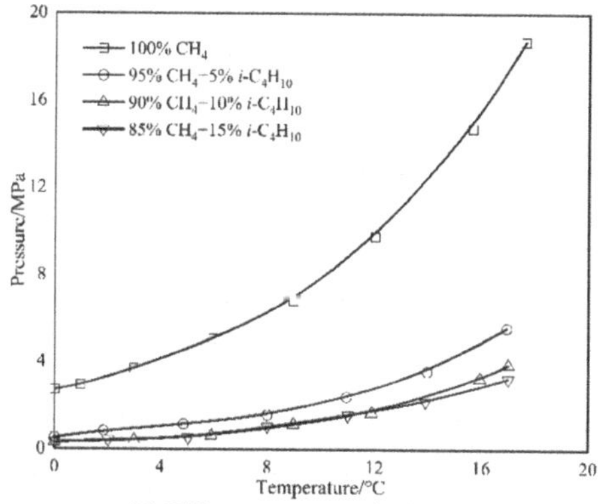

(c) Different contents of n-butane

Fig. 3.12 Impact of different contents of ethane, propane, and n-butane on hydrate formation

at different natural gas density are shown in Fig. 3.13, illustrating for the same components of natural gas, the higher pressure means the higher critical temperature for hydrate formation, and vice versa. Besides, the higher density ratio of natural gas means the higher critical temperature for hydrate formation under the same pressure, and no hydrate would form if the temperature exceeds the critical temperature.

3.4.3 Thermodynamic Inhibitor

The impact of the thermodynamic inhibitor on the hydrate phase equilibrium curve is shown in Fig. 3.14, illustrating the hydrate phase equilibrium curve would move to the left. This is because the thermodynamic equilibrium between water and gas molecules, and the chemical potential of aqueous solutions or hydrates are changed. Here, it is more difficult for hydrate to form and stabilize under the same temperature and pressure.

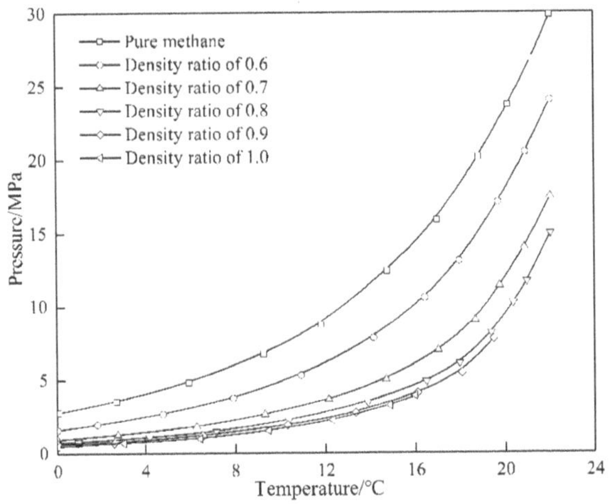

Fig. 3.13 Hydrates phase equilibrium curve at different relative density

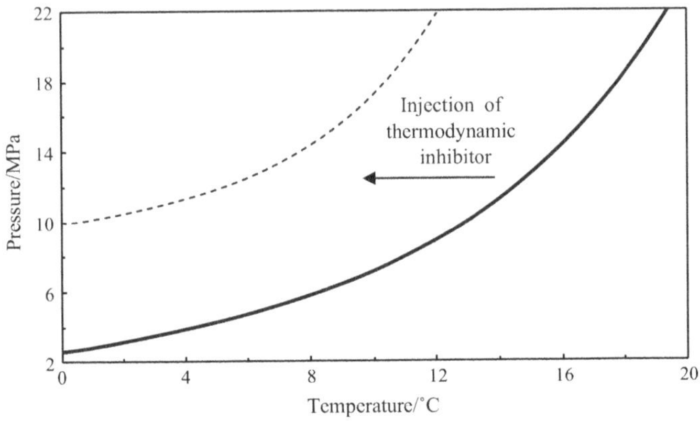

Fig. 3.14 Impact of thermodynamic inhibitor on hydrate phase equilibrium curve

3.4.4 Sand Content

To identify the impact of quartz sand in fluid on the thermodynamics and kinetics laws of hydrate formation, quartz sands are added in the high-pressure reactor to simulate the sand production and porous media in deepwater well. In the whisking system of quartz sands, hydrate phase equilibrium curves with different sizes of the sand particle are tested, whose testing program is shown in Table 3.4. The agitated high-pressure test device (Fig. 3.15) is designed by Hydrate Research Center of China University of Petroleum (East China), which is mainly composed of the reaction vessel with a volume of 1 L, electromagnetic whisk, water bath, the system of gas

Table 3.4 Test scheme

Types of quartz sands deposition layer	Particle size of quartz sands/μm	Temperature/K	Pressure/MPa
1	96–180	281.6–284.3	6.0–8.0
2	180–380	281.1–284.2	5.8–7.8
3	212–380	274.7–282.8	3.3–6.7
4	380–830	275.0–282.7	3.2-6.8

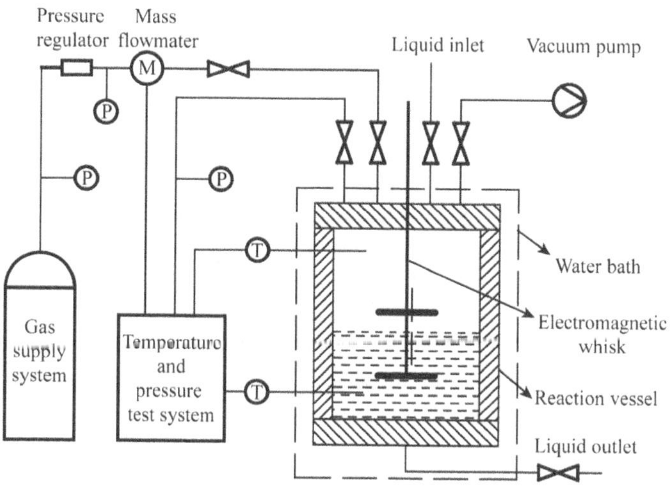

Fig. 3.15 High-pressure whisked device for hydrate formation

supply and testing the temperature and pressure and so on. The testing steps are as follows:

(1) Wash sands
Wash quartz sands with distilled water and dry it for later use;
(2) Fill sands
Fill quartz sands into the reaction vessel according to the weight ratio of sand to distilled water (1:3);
(3) Pump vacuum
(4) Inject gas and keep the pressure stabilized
The reaction vessel stands for 12 h after gas injection to get full diffusion and dissolution of methane and to keep the pressure stabilized;
(5) Start to form the hydrate
Switch on the whisks and set low temperature, hydrate will start to form under large subcooling degree;
(6) Start to decompose the hydrate
Increase the temperature gradually to carry out the hydrate decomposition and record its variation of temperature and pressure;

(7) Repeat the steps above
Repeat the steps above by changing the vessel pressure to obtain multiple hydrate phase equilibrium points.

In the whisking system of quartz sands and pure water system, the methane hydrate phase equilibrium curves (Fig. 3.16) show the phase equilibrium points of methane hydrate in whisking system are consistent with those in pure water system obtained by Deaton and Frost [21], Nakamura et al. [22] and Adisasmito et al. [23]. Besides, if the particle size of quartz sands is larger than 96 μm, we could neglect its influence on the hydrate phase equilibrium.

Compared with the whisking system of quartz sands, the sand-filling system could better reflect the influence of sand particles on the hydrate phase equilibrium points. Therefore, Anderson and Tohidi used a nanometer-particle of sands to carry out the hydrate formation experiment in a sand-filling system. The experiment result showed it was more difficult for hydrate to form in the sand-filling system with a pore radius of less than 50 nm. Particularly, under the same pressure, the temperature for hydrate formation in the sand-filling system is about 1 °C less than that in a pure water system. Also, smaller sand size or pore radius is unfavorable for hydrate nucleation possibly owing to the fact that the capillary force of pore could increase the injection pressure of water and change the water activity in the porous medium. However, Turner's [24] proved that the capillary force does not influence the hydrate formation when the pore radius is greater than 60 nm in the sand-filling system. And based on the hydrate phase equilibrium data, Chen [25] calculated and believed the critical value of pore radius is 58.68 nm.

Fig. 3.16 Methane hydrate phase equilibrium curves in whisking system of quartz sands and pure water system

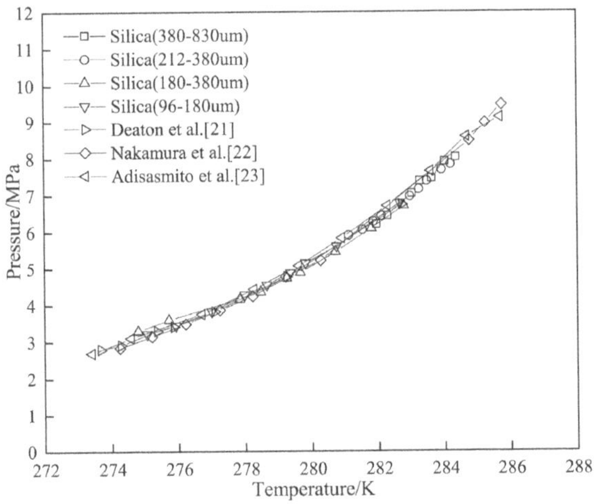

3.5 Prediction and Influence Factors of Hydrate Formation Area in Deepwater Gas Well

3.5.1 Prediction Methods of Hydrate Formation Area Under Different Working Conditions

The deepwater gas exploitation mainly includes several operation processes such as drilling, well control, testing, production. And prediction methods of hydrae formation area under different working conditions are similar. Comprehensively considering the temperature field distribution in well/pipeline and hydrate phase equilibrium curve, Wang et al. and Sun et al. [14, 26–29] proposed one prediction method of hydrate formation area in the deepwater gas well. Diagram of hydrate formation area in well and pipeline is shown in Figs. 3.17 and 3.18. The prediction process are as follows:

(1) Numerically solve the temperature field in well and temperature and pressure field in the pipeline under different working conditions;
(2) Solve the hydrate phase equilibrium curve in the traditional form of temperature-pressure according to hydrate phase equilibrium equation;
(3) For well, convert the temperature-pressure curve into the hydrate phase equilibrium curve in the form of temperature-depth;
(4) Draw the curve of well temperature field and the hydrate phase equilibrium curve in the form of temperature-depth pari passu; Similarly, for the pipeline, draw the curve of temperature and pressure field and the hydrate phase equilibrium curve in the form of temperature-depth. Afterwards, the intersection area of two curves refers to the hydrate formation area.

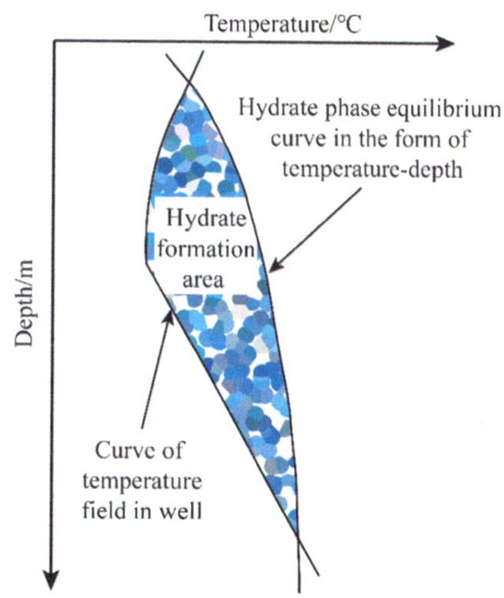

Fig. 3.17 Diagram of hydrate formation area in well

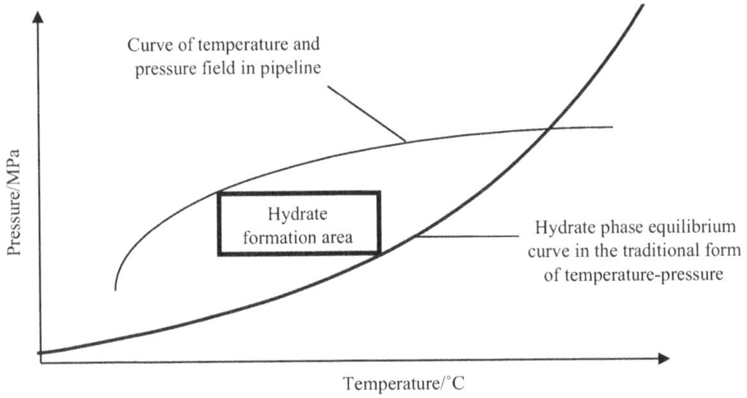

Fig. 3.18 Diagram of hydrate formation area in pipeline

Note that the greater length of the intersection area in longitudinal direction means the larger hydrate formation area; conversely, the greater width in transverse direction means the greater subcooling degree for hydrate formation, which indicates the faster rate of hydrate formation.

3.5.2 *Influencing Factors of Hydrate Formation Area During Drilling*

Taking one deepwater well as an example, the influence of parameters such as circulation flow of drilling fluid, inlet temperature of drilling fluid and inhibitor concentration on hydrate formation area during drilling are analyzed. Basic parameters of the deepwater well include water depth of 1500 m, casing down to 3500 m below the water surface, drilling tools of 127 mm, drill bit of 215.9 mm, choke pipeline inner diameter of 76.2 mm, volumetric flow rate of 54 L/s, seabed temperature of 2 °C, stratum temperature gradient of 3 °C/100 m and the equivalent density of stratum rupture with 1.27 g/cm^3.

3.5.2.1 Circulation Flow of Drilling Fluid

Through the above prediction method, hydrate formation area under different circulation flows of drilling fluid is calculated and shown in Fig. 3.19 [30]. As the velocity of circulation flow decreases, heat exchange between well fluid and external environment would last for a long period. So, the well annulus temperature decreases above the mudline (1500 m) but increases near the well bottom under a low temperature of seawater and high temperature of the stratum. This is to say, the well annulus temperature tends to be close to the external environment temperature, which gradually

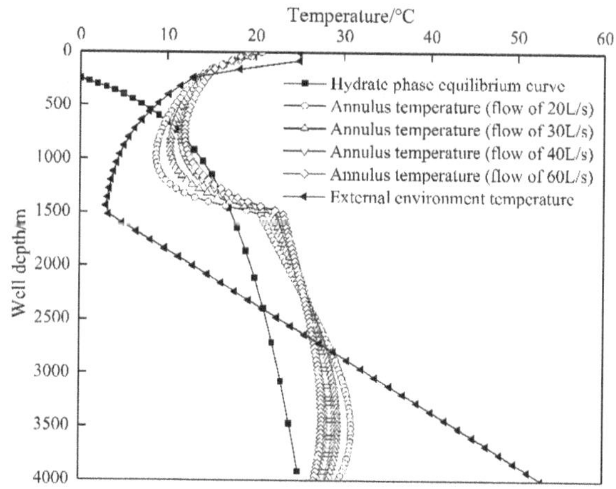

Fig. 3.19 Hydrate formation area in different circulation flows of drilling fluid

enlarges the subcooling degree and hydrate formation area. Furthermore, hydrates could be likely to gather and form near the blowout preventer, so we could properly increase the circulation flow to narrow the hydrate formation area, finally reduce the hydrate blockage risks of pipeline.

3.5.2.2 Inlet Temperature of Drilling Fluid

The hydrate formation area under different inlet temperatures of drilling fluid is calculated and shown in Fig. 3.20 [14], showing the inlet temperature of drilling fluid significantly affects the well annulus temperature distribution in the normal cyclic state, and the hydrate formation area would gradually become small with the inlet temperature until there is no hydrate formation area at 30 °C. Thus, hydrate

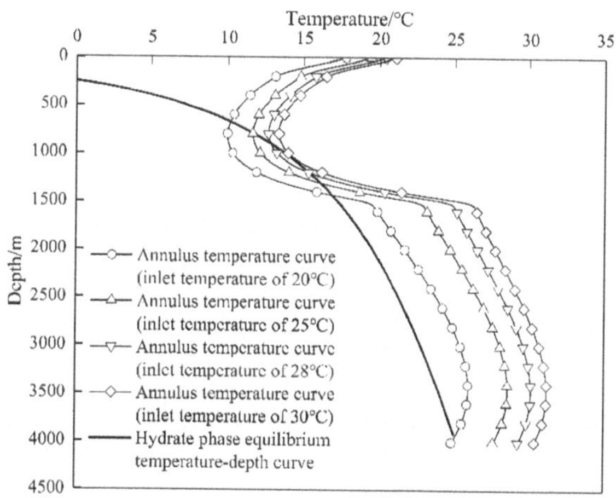

Fig. 3.20 Hydrate formation area under different inlet temperatures of drilling fluid

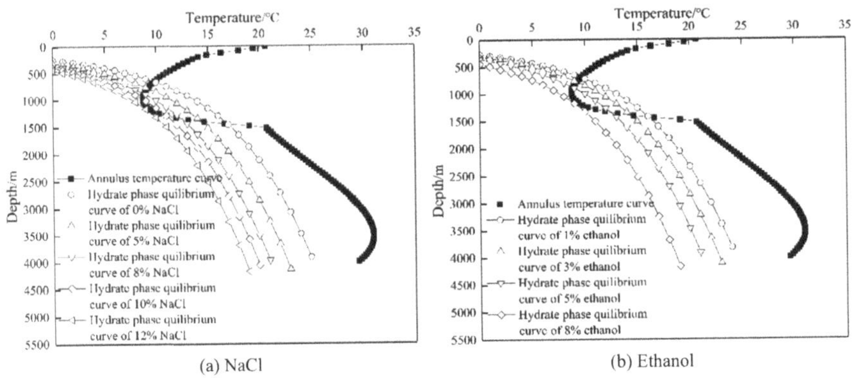

Fig. 3.21 Hydrate formation area under different inhibitor concentrations

blockage risks, in a manner, could be reduced by taking thermal insulation measures for the flow-back drilling fluid.

3.5.2.3 Inhibitor Concentration

Inhibitor injection is the most popular method for hydrate prevention and management during drilling. The hydrates formation area under different concentrations of NaCl and ethanol is shown in Fig. 3.21 [14], illustrating the hydrate phase equilibrium curves would move to the left with the concentration of NaCl and ethanol, leading to the smaller hydrate formation area. Besides, the critical concentration of NaCl and ethanol without hydrate formation is 12% and 8% respectively, indicating ethanol exhibits a stronger inhibitory effect against hydrate formation than NaCl.

3.5.3 Influencing Factors of Hydrate Formation Area During Well Control

The influence of off production time and choke pipeline inner diameter on the hydrate formation area during well control is calculated and analyzed based on relevant parameters in the deepwater well seen in Sect. 3.5.2 [8].

3.5.3.1 Off Production Time

The hydrate formation area under different off production time is shown in Fig. 3.22, illustrating the subcooling degree and hydrate formation area would enlarge gradually with off production time. This is because that the longer off production time means

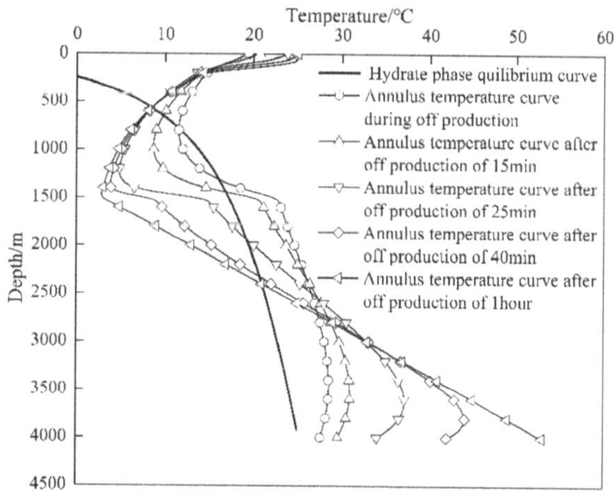

Fig. 3.22 Hydrate formation area under different off production time

the more sufficient heat exchange between well annulus fluid and external environment, and the well annulus temperature is roughly equal to the low-temperature environment when well shut-in lasts for 1 h. Notice that the longer off production time means the lower temperature at mudline, hydrate formation will easily be found in the pipeline near the blowout preventer. Therefore, off production time should be minimized to prevent hydrate formation during well control.

3.5.3.2 Inner Diameter of Choke Pipeline

The hydrate formation area under different inner diameters of the choke pipeline is shown in Fig. 3.23. The long choke pipeline and its small size cause the large frictional resistance in the choke pipeline, which further increase the well annulus

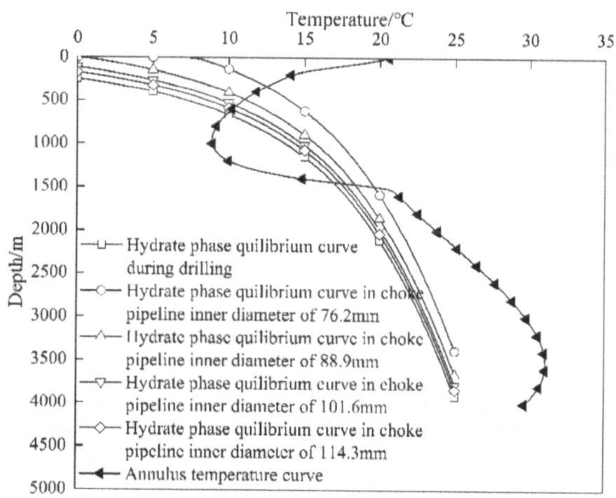

Fig. 3.23 Hydrate formation area under different inner diameters of choke pipeline

pressure and shift the hydrate phase equilibrium curve during well control. It can be seen from the figure that as the inner diameter of the choke pipeline decreases, the hydrate phase equilibrium curve would move to the top right, leading to the larger subcooling degree and hydrate formation area. Thus, the appropriate inner diameter of the choke pipeline should be chosen to narrow the hydrate formation area during well control.

3.5.4 Influencing Factors of Hydrate Formation Area During Well Test

Based on one deepwater well, the influence of parameters such as gas production, geothermal gradient and seawater depth on hydrate formation area is analyzed during well testing. Basic parameters of this selected well include the sea surface temperature of 25 °C, bottom-hole pressure of 28 MPa, geothermal gradient of 1.77 °C/100 m, seawater depth of 768 m, well depth of 1942.59 m.

3.5.4.1 Gas Production

The hydrate formation area when gas production changes from 1×10^4 m^3/d to 150×10^4 m^3/d is shown in Fig. 3.24. At one well depth, the well temperature increases with gas production. Besides, a larger hydrate formation area would be obtained if gas production is less than 15×10^4 m^3/d. Hydrate would be only found near the wellhead if gas production reaches 15×10^4 m^3/d. No hydrate would be found in well when gas production is greater than 20×10^4 m^3/d. This is because the lower gas production means higher wellhead pressure at certain temperature and pressure of well bottom, meanwhile results in the sufficient heat exchange between well fluid and low-temperature environment. Thus, the fluid temperature becomes lower and hydrate is easier to form. Affected by the high-temperature reservoir, the fluid temperature is far higher than the phase equilibrium temperature of hydrate under high gas production, so no hydrate could be found in well.

3.5.4.2 Geothermal Gradient

The hydrate formation area under different geothermal gradients is shown in Fig. 3.25 [28], illustrating the hydrate formation area gradually diminishes with the geothermal gradient. Specifically, the hydrate formation area is 420–1400 m and 500–1090 m under the respective geothermal gradients of 1.0 °C/100 m and 1.5 °C/100 m, and no hydrate would be found in well when the geothermal gradient is higher than 2.0 °C/100 m, because the high geothermal gradient or well temperature is not

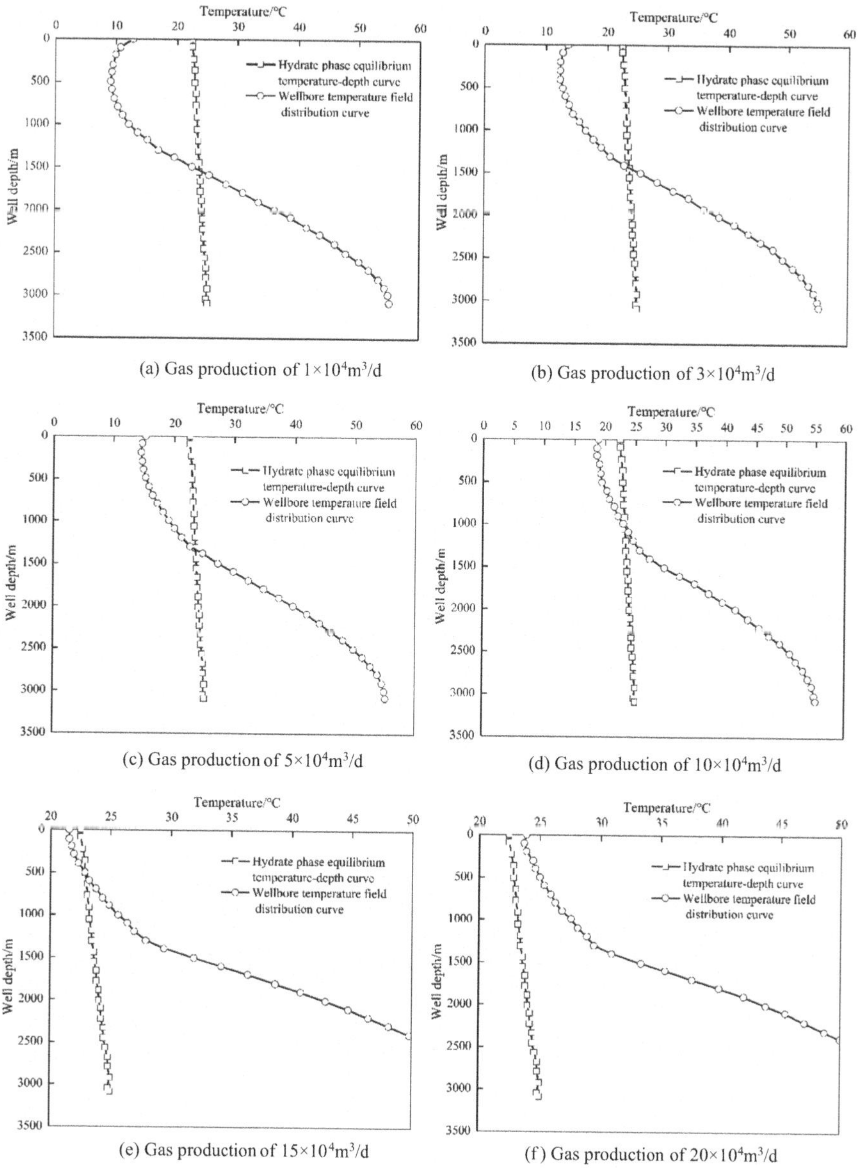

Fig. 3.24 Hydrate formation area under different gas production

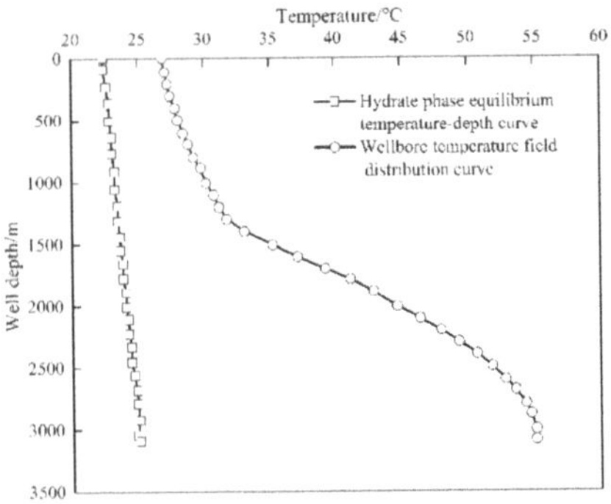

(g) Gas production of $30\times10^4m^3/d$

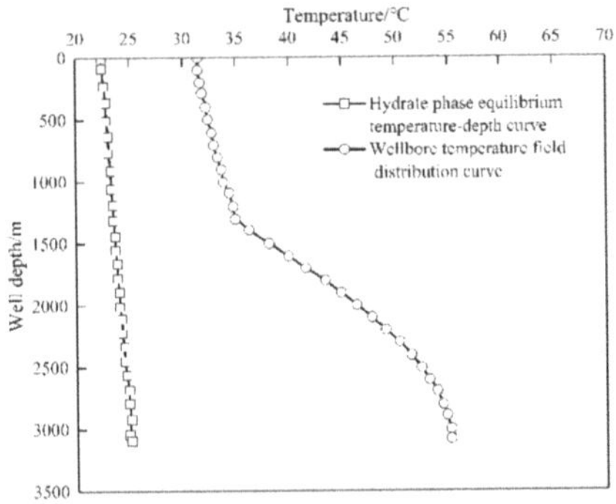

(h) Gas production of $50\times10^4m^3/d$

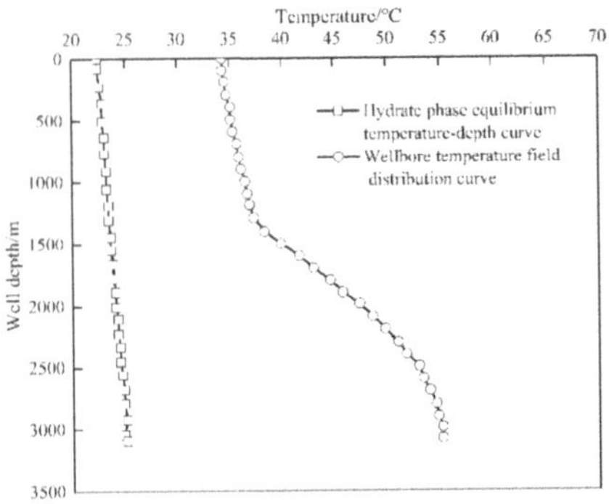

(i) Gas production of $70\times10^4m^3/d$

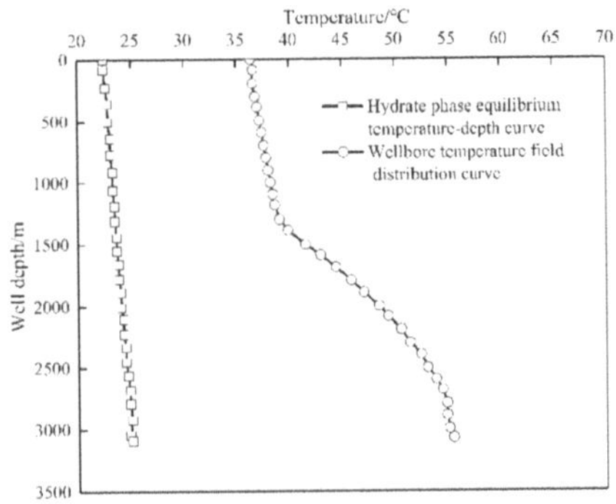

(j) Gas production of $90\times10^4m^3/d$

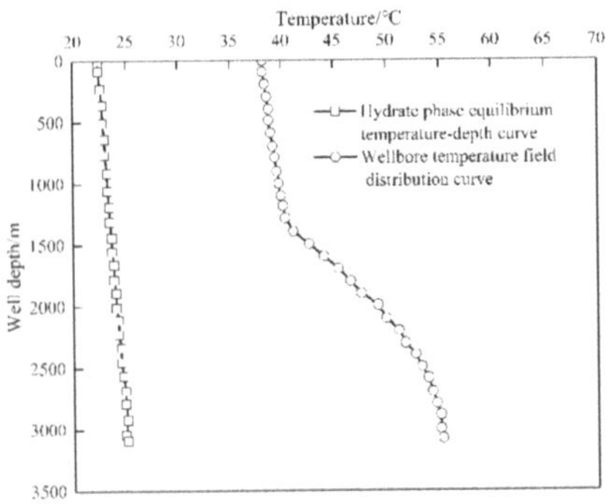

(k) Gas production of $110\times10^4m^3/d$

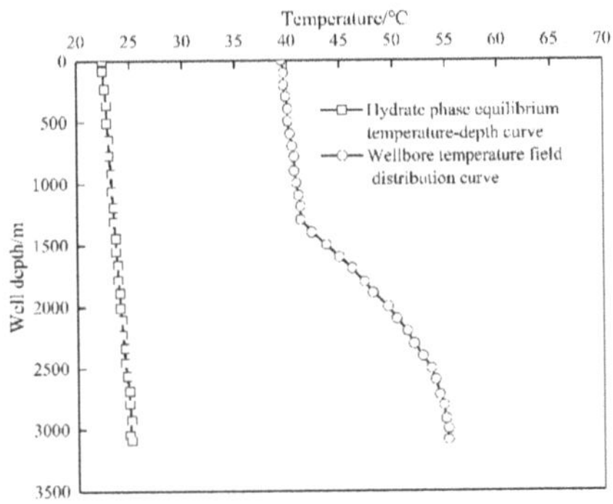

(l) Gas production of $130\times10^4m^3/d$

Fig. 3.24 (continued)

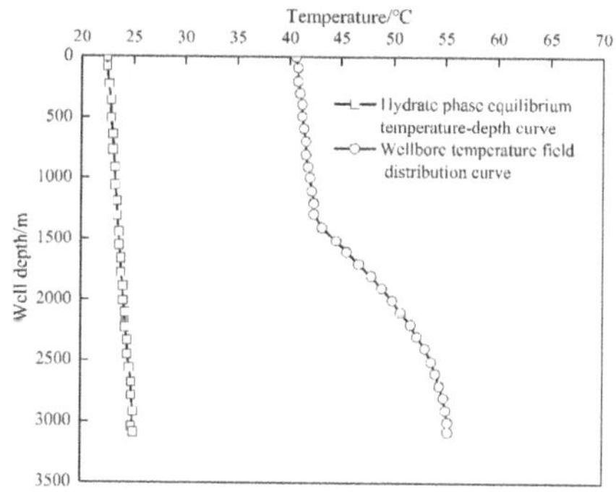

(m) Gas production of $150\times10^4 m^3/d$

Fig. 3.24 (continued)

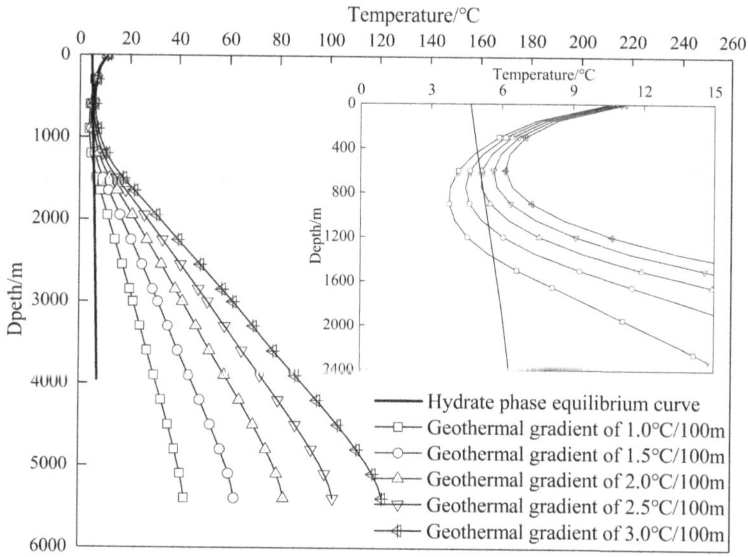

Fig. 3.25 Hydrate formation area in different temperature gradients

conducive to the hydrate formation. Therefore, we should pay attention to the hydrate formation under small geothermal gradients during testing of the deepwater gas well.

3.5.4.3 Seawater Depth

The hydrate formation area under different seawater depths is shown in Fig. 3.26 [28], which shows that the hydrate formation area gradually enlarges with the seawater depth. Specifically, the hydrate formation area at a water depth of 1000 m and 750 m is

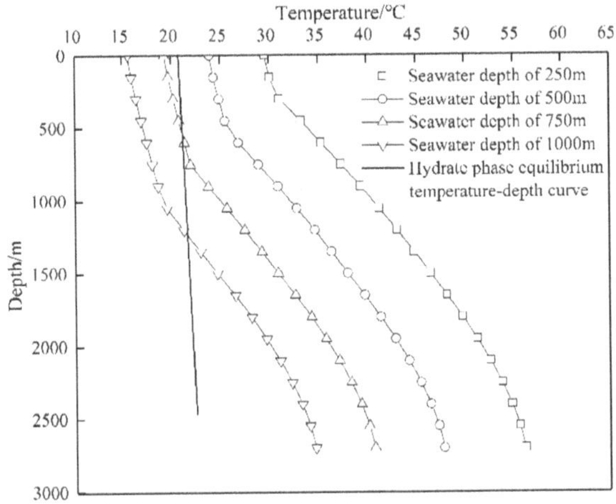

Fig. 3.26 Hydrate formation area under different water depths

0–1224 m and 0–623 m respectively, and no hydrate would be found at a water depth of 500 m and 250 m. It indicates that the well temperature near mudline becomes lower with the seawater depth, resulting in the enlargement of the hydrate formation area.

References

1. Reyna EM, Stewart SR (2001) Case history of the removal of a hydrate plug formed during deep water well testing. In: SPE/IADC drilling conference. Society of Petroleum Engineers
2. Arrieta VV, Torralba AO, Hernandez PC et al (2011) Case history: lessons learned from retrieval of coiled tubing stuck by massive hydrate plug when well testing in an ultradeepwater gas well in Mexico. SPE Prod Oper 26(4):337–342
3. Trummer SA, Mohallem R, Franco E et al (2013) Hydrate remediation during well testing operations in the deepwater Campos Basin, Brazil. In: SPE/ICoTA coiled tubing & well intervention conference & exhibition. Society of Petroleum Engineers
4. Wang Z, Sun B, Gao Y et al (2008) Simulation computation of well kill ing with deepwater driller's method. Acta Pet Sin 05:786–790
5. Zhang Z, Sun B, Wang Z et al (2015) Prediction and analysis of natural gas hydrate formation region during deep water gas well testing. Chin J Hydrodyn 30(2):167–172
6. Zhao Y, Wang Z, Sun B et al (2015) Two-phase flow heat transfer modeling for prediction of gas hydrate formation region in deepwater gas-well with water content. In: The 27th national hydrodynamics symposium
7. Wang Z, Zhao Y, Sun B et al (2016) Heat transfer model for annular-mist flow and its application in hydrate formation risk analysis during deepwater gas well testing. Chin J Hydrodyn 31(01):20–27
8. Wang Z, Sun B, Cheng H et al (2009) Prediction of natural gas hydrate formation area during deep water well control. Chin J Appl Mech 26(02):224–229
9. Wang Z, Yu J, Zhang J et al (2018) Improved thermal model considering hydrate formation and deposition in gas-dominated systems with free water. Fuel 236:870–879

10. Wang Z, Liao Y, Zhang W et al (2018) Coupled temperature field model of gas-hydrate formation for thermal fluid fracturing. Appl Therm Eng 133:160–169
11. Wang Z, Zhao Y, Sun B et al (2016) Modeling of hydrate blockage in gas-dominated systems. Energy Fuels 30(6):4653–4666
12. Sun X, Sun B, Wang Z et al (2018) A hydrate shell growth model in bubble flow of water-dominated system considering intrinsic kinetics, mass and heat transfer mechanisms. Int J Heat Mass Transf 117:940–950
13. Sawant P, Ishii M, Mori M (2009) Prediction of amount of entrained droplets in vertical annular two-phase flow. Int J Heat Fluid Flow 30(4):715–728
14. Wang ZY, Sun BJ, Cheng HQ et al (2008) Prediction of gas hydrate formation region in the wellbore of deepwater drilling. Pet Explor Dev 35(6):731–735
15. Wang Z, Sun B, Gao Y et al (2010) Study on well annulus multiphase flow characteristic of gas kick during hydrate reservoir drilling. J Basic Sci Eng 18(01):129–140
16. Lorenzo MD, Aman ZM, Kozielski K et al (2014) Underinhibited hydrate formation and transport investigated using a single-pass gas-dominant flowloop. Energy Fuels 28(11):7274–7284
17. Gao Y, Sun B, Wang Z et al (2008) Calculation and analysis of wellbore temperature field in deepwater drilling. J China Univ Pet (Edition of Natural Science) 32(02): 58–62
18. Alves IN, Alhanati FJS, Shoham O (1992) A unified model for predicting flowing temperature distribution in wellbores and pipelines. SPE Prod Eng 04(7):363–367
19. Hasan AR, Kabir CS, Sayarpour M (2010) Simplified two-phase flow modeling in wellbores. J Petrol Sci Eng 72(1):42–49
20. Zhang J, Wang Z, Liu S et al (2019) Prediction of hydrate deposition in pipelines to improve gas transportation efficiency and safety. Appl Energy 253.113521
21. Deaton WM, Frost EM (1949) Gas hydrates and their relation to the operation of natural-gas pipe lines. American Gas Association
22. Nakamura T, Makino T, Sugahara T et al (2003) Stability boundaries of gas hydrates helped by methane—structure-H hydrates of methylcyclohexane and cis -1,2-dimethylcyclohexane. Chem Eng Sci 58(2):269–273
23. Adisasmito S, Frank RJ, Sloan ED (1991) Hydrates of carbon dioxide and methane mixtures. J Chem Eng Data 36(1):68–71
24. Turner D, Boxall J, Yang S et al (2005) Development of a hydrate kinetic model and its incorporation into the OLGA2000® transient multiphase flow simulator. In: 5th international conference on gas hydrates, Trondheim, Norway 12–16
25. Chen Q, Ye S, Liu C et al (2007) Research on formation kinetics of methane hydrate in porous media. Mar Geol Quat Geol 01:111–116
26. Wang Z, Sun B, Wang X et al (2014) Prediction of natural gas hydrate formation region in wellbore during deep-water gas well testing. J Hydrodyn Ser. B 26(4): 568–576
27. Zhang J, Wang Z, Sun B et al (2019) An integrated prediction model of hydrate blockage formation in deep-water gas wells. Int J Heat Mass Transf 140:187–202
28. Wang Z, Sun B, Wang X et al (2014) Prediction of natural gas hydrate formation region in wellbore during deep-water gas well testing. J Hydrodyn 26(4):568–576
29. Wang Z, Zhao Y, Zhang J et al (2018) Flow assurance during deepwater gas well testing: Hydrate blockage prediction and prevention. J Petrol Sci Eng 163:211–216
30. Wang Z, Sun B (2009) Annular multiphase flow behavior during deep water drilling and the effect of hydrate phase transition. Pet Sci 6(1):57–63

Chapter 4
Influence of Hydrate Phase Transition on Multiphase Flow in Deepwater Gas Well

During drilling, as the gas invasion and fluid migration under high-pressure and low-temperature environments, hydrate formation and decomposition have occurred successively. Such a hydrate phase transition would affect the gas invasion laws and overflow characteristics. In this chapter, the influence of hydrate phase transition on the rheology of drilling fluid is introduced firstly; then the well annulus multiphase flow model during drilling is established, and influence of hydrate phase transition on bubbles migration is analyzed; the influence of hydrate phase transition on well multiphase flow with inhibitors and without inhibitors is discussed at last.

4.1 Influence of Hydrate Phase Transition on the Rheology of Drilling Fluid

For the influence of the hydrate phase transition on the rheology of drilling fluid, many studies have been carried out [1–5]. Peysson [6] believed that hydrate particles and fluids could be regarded as certain phases with new friction factor. That being said, hydrate particles would increase the friction factor between them and pipewall, causing the pressure drop to increase; Pei et al. [7] studied the rheological properties in the water-in-oil emulsion system after hydrate formation, finding that the subcooling degree had a significant influence on the system viscosity; Webb et al. [8, 9] studied the rheological properties in the water-in-oil system by high-pressure rheometer and discussed the influence of time, shear rate, water content and temperature on its rheological properties. Wang et al. and Sun et al. [10–12] further discussed the influence of the hydrate phase transition on the rheology of drilling fluid by constructing a set of device for hydrate multiphase flow experiments, which is mainly introduced in this section.

© Science Press and Springer Nature Singapore Pte Ltd. 2020

Z. Wang et al., *Natural Gas Hydrate Management in Deepwater Gas Well*,
https://doi.org/10.1007/978-981-15-6418-5_4

4.1.1 Experimental Device

The device used for multiphase flow experiment is shown in Fig. 4.1 [10, 11], where the total volume is 9.602×10^{-3} m^3 and the test section is mainly composed of three horizontal pipes (stainless steel): pipe No. 1, pipe No. 2 and pipe No. 3 with a respective length of 3.07 m, 0.76 m and 4.43 m, an inner diameter of 0.0254 m and an outer diameter of 0.0635 m. One transparent PVC pipe (length of 0.3 m and an inner diameter of 0.0254 m) for observation is installed between pipe No. 1 and pipe No. 2 and the high-pressure hose is used to connect pipes and centrifugal pump. Besides, all the pipes and centrifugal pumps are coated with thermal insulation materials to avoid heat loss. The pressure and temperature difference between pipes are measured by differential pressure transmitter and temperature sensor respectively.

The multiphase flow experiment is carried out in bubble flow with a void fraction of 4.5%. The initial subcooling degree is determined as 6.0 K to reduce experiment error, and the methane with a purity of 99.99% is used as the experimental gas. After vacuum pumping, methane gas is injected into the flow loop through a gas cylinder. Meantime, the electromagnetic flowmeter is used to measure the volume flow of gas. When the system temperature stabilizes at one target value, deionized water is delivered into the flow loop by a plunger pump, then hydrate begins to form. During the whole process, the fluid transport and flow pattern changes in the PVC pipe are recorded by a high-speed camera. Besides, the temperature and pressure, pressure drop and flow rate in each time interval are constantly monitored and acquired by the computer.

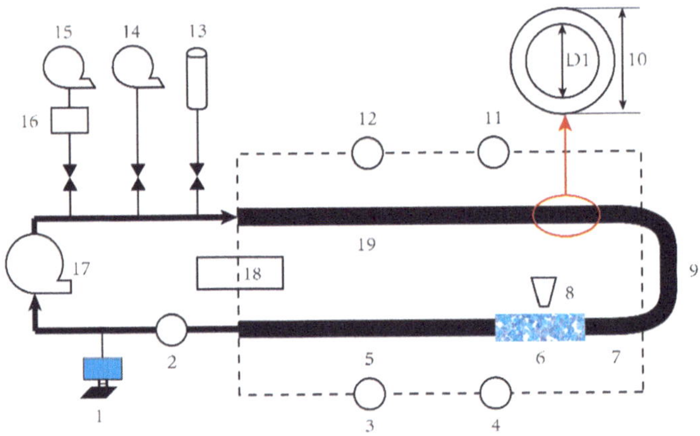

Fig. 4.1 Multiphase flow experimental device, 1—Computer, 2—Electromagnetic flowmeter, 3—Temperature sensor, 4—Differential pressure transmitter, 5—Horizontal pipe 1, 6—PVC transparent pipe, 7—Horizontal pipe 2, 8—High-speed camera, 9—Horizontal pipe 3, 10—Outer diameter of pipe, 11—Differential pressure transmitter, 12—Temperature sensor, 13—Gas cylinder, 14—Plunger pump, 15—Vacuum pump, 16—Surge flask, 17—Electric pump, 18—Cooler, 19—Horizontal pipe 4

Xanthan gum aqueous solution (simply as "XG solution"), one type of non-Newtonian fluid and do not ignore the influence of fluid rheology on mass transfer, is used as a carrier fluid in the multiphase flow experiment. In the multiphase flow system, the mass transfer coefficient and interface area mainly depend on the rheological properties of carrier fluid. Notice the interface area is difficult to be measured accurately, so the volumetric mass transfer coefficient is often used for calculation [13–15].

4.1.2 Drilling Fluid Rheology Model Considering Hydrate Formation

The flow resistance could be reduced by sufficient agitation during the multiphase flow experiment, and the migration process of gas molecules from the liquid phase to the hydrate phase could be neglected. Thus, hydrate formation in the water continuous phase is deemed as the diffusion process of methane molecules from the gas phase to the water phase [11]. The driving force for hydrate formation is derived from methane concentration difference between the gas-water two-phase interface and water phase under hydrate phase equilibrium condition. So, the mass flux of methane molecules could be calculated by Eq. 4.1.

$$J_{CH_4}=k_{CH_4}\left(C_I-C_{eq}\right) \tag{4.1}$$

where J_{CH_4} shows the mass flux of methane molecules, kg/(m^2 s); k_{CH_4} refers to the mass transfer coefficient of methane, m/s, calculated by Sherwood number; C_I is the methane concentration in the gas-water interface, kg/m^3; C_{eq} is the methane concentration in the water phase under hydrate phase equilibrium condition, kg/m^3.

Hydrate formation is related to the surface area between the gas and liquid phases [16]. The driving force for hydrate formation could be simplified as the molar fraction difference of methane between gas-water interface and water phase, so the formation rate of methane hydrate could be shown in Eq. 4.2 [17].

$$\frac{dn_{CH_4}}{dt}=\frac{dn_H}{dt}=\frac{1}{M_{CH_4}}k_{CH_4}\alpha V_{tot}\left(C_I-C_{eq}\right)=\frac{1}{M_{CH_4}}k_{CH_4}\alpha V_{tot}C_w\left(x_I-x_{eq}\right) \tag{4.2}$$

where $\frac{dn_{CH_4}}{dt}$ represents the mass transfer rate of methane molecules, mol/s; $\frac{dn_H}{dt}$ represents the methane consumption rate due to hydrate formation, mol/s; M_{CH_4} is the molar mass of methane, kg/mol; V_{tot} stands for the total volume of the system, m^3; α is the concentration of specific area, m^{-1}; x_I shows the gas molar fraction in the gas-water interface, dimensionless; x_{eq} shows the gas molar fraction in water under hydrate phase equilibrium condition, dimensionless.

Nakanoh and Yoshida [18] inferred the Sherwood number of non-Newtonian fluid could be calculated by Eq. 4.3:

$$Sh = 12C\frac{1}{\sqrt{\pi}}\sqrt{1.07}n^{1/3}Sc^{0.5}Re^{\frac{2n}{2(1+n)}}Fr^{\frac{11n-4}{30(1+n)}}Bo^{0.6} \tag{4.3}$$

where C stands for the empirical constant of Sherwood Number; n is the non-Newtonian index of XG solution; Sh shows the Sherwood Number; Sc shows the Schmidt Number; Re shows the Reynolds Number; Fr shows the Froude Number; Bo shows the bonding Number, representing the interfacial tension influence of gas and non-Newtonian fluid.

The interfacial tension is generally sensitive to ambient temperature, and it is also affected by additives concentration for non-Newtonian fluid. Muthamizhi et al. [19] studied the interfacial tension of XG solution with a concentration of 0.1–0.6%, and considered the interfacial tension as a function of ambient temperature and solution concentration. The concentration of XG solution in this experiment is 0.1–0.3%, and its interfacial tension could be calculated as:

$$\begin{aligned}\sigma_{XG} &= 0.860052 - 0.00473T_{sys} - 0.03529C_{XG} + 0.00015T_{sys}C_{XG} \\ &\quad + 0.00000705T_{sys}^2 - 0.00345C_{XG}^2\end{aligned} \tag{4.4}$$

where σ_{XG} represents the interfacial tension of XG solution, N/m; T_{sys} is the ambient temperature, K; C_{XG} refers to the concentration of XG solution, %.

The rheological properties of the XG solution could be evaluated by a capillary-gravity viscometer. Shear stress in the non-Newtonian fluid could be expressed as the function of n' index, K' coefficient and flow rate, which is shown in Eq. 4.5. Besides, the shear rate in the non-Newtonian fluid is the function of flow velocity and n' index is shown in Eq. 4.6. Therefore, shear stress in the non-Newtonian fluid can further represent a function of yield stress, stress rate, n' index and K' coefficient and be shown in Eq. 4.7 [20].

$$\tau_{XG} = \frac{D}{4}\frac{\Delta P}{L} = K'\left(\frac{8\nu_{XG}}{D}\right)^{n'} \tag{4.5}$$

$$\gamma_{XG} = \frac{8\nu_{XG}}{D}\frac{3n'+1}{4n'} \tag{4.6}$$

$$\tau_{XG} = \tau_0 + K'\gamma_{XG}^{n'} \tag{4.7}$$

where ΔP shows the pressure drop in a horizontal pipe, Pa; L is the pipe length, m; ν_{XG} refers to the flow rate of XG solution, m/s; n' stands for the non-Newtonian behavior index; K' stands for the consistency coefficient; τ_0 shows the yield stress, Pa; τ_{XG} shows the shear stress of XG solution, Pa; γ_{XG} refers to the shear rate of XG solution, s^{-1}.

When the concentration of XG solution increases from 0.1% to 0.3%, the relation between shear rate and shear stress of XG solution is shown in Fig. 4.2, which shows the shear stress increases logarithmically with shear rate, and exhibits greater logarithmic with XG solution concentration. In addition, all the regression lines of experimental data start from the origin, indicating the yield stress of τ_0 is zero. Therefore, the apparent viscosity of XG solution could be expressed as

$$\mu_{XG} = \frac{\tau_{XG}}{\gamma_{XG}} = K'\gamma_{XG}^{n'-1} \tag{4.8}$$

The relations between n' index, consistency coefficient K' and XG solution concentration are shown in Fig. 4.3. The exponential regression for n' and K' is performed as below:

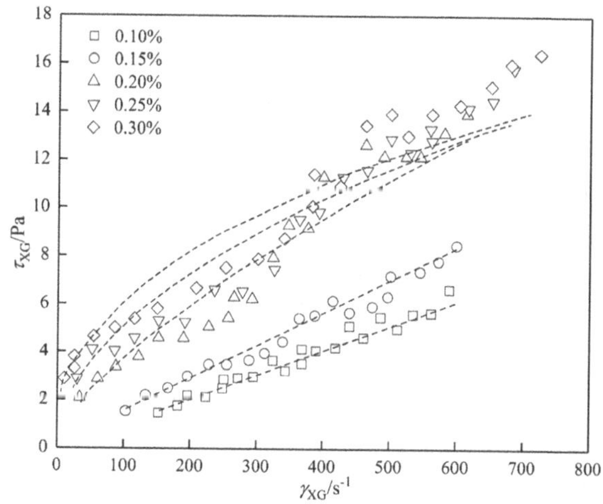

Fig. 4.2 Relation between shear stress and shear rate

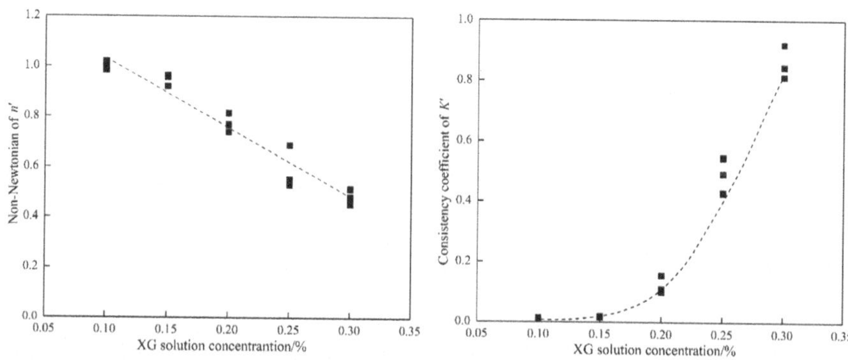

(a) (b)

Fig. 4.3 Relation between n' index **a**, K' coefficient **b** and XG solution concentration

$$n' = -265.65C_{XG} + 1.2831 \tag{4.9}$$

$$K' = 2 \times 10^{10} C_{XG}^{4.094} \tag{4.10}$$

Substituting Eqs. 4.9 and 4.10 into Eq. 4.8 could obtain the ultimate expression about apparent viscosity of XG solution:

$$\mu_{XG} = 2 \times 10^{10} C_{XG}^{4.094} \gamma_{XG}^{-265.65C_{XG}+0.2831} \tag{4.11}$$

In particular, the empirical rheology model of drilling fluid works well only in certain experimental range (XG solution concentration of 0.1%–0.3%, shear rate of 2.94–723 s^{-1}).

4.1.3 Hydrate Formation Integration Constant of C_{HF}

The initial subcooling degree is close to 6.0 K, but 0 K at the later stage of hydrate formation. Under the flow rate of 1.18 m/s in flow loop, methane hydrate formation rate in the whole process of hydrate formation is shown in Fig. 4.4a. it could be found that the hydrate formation rate would drop sharply in the initial period and then remain essentially unchanged when the concentration of XG solution is 0% and 0.10%, while the hydrate formation rate would increase sharply in the final period. This indicates the process of hydrate formation in multiphase flow is not completely controlled by a subcooling degree.

In the initial period, hydrate shell forms on the bubble surface [21, 22], and then grow inwards and outwards to prevent further contact of gas and liquid phase, so hydrate formation is inhibited. In the middle period, bubbles are collided with

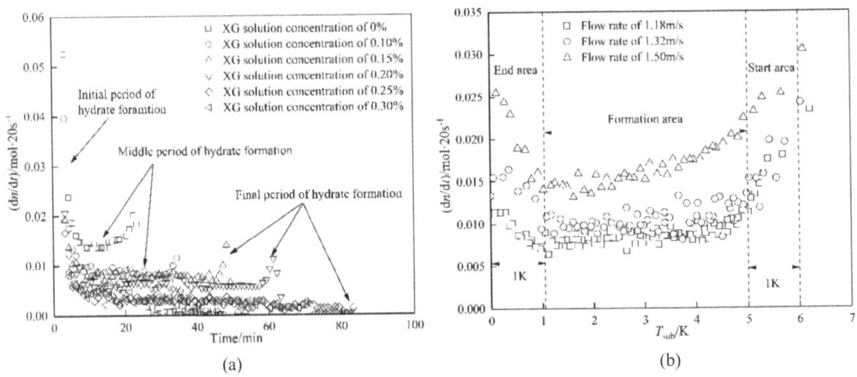

Fig. 4.4 The whole process of methane hydrate formation, **a** influence of XG solution concentration, **b** influence of XG solution flow rate

each other and ruptured, which increase the gas-liquid interface area. Thus, hydrate reformation on the bubble surface would occur and hydrate formation rate increases. Meanwhile, both the inhibition action caused by extended growth of hydrate shell and the promotion effect caused by bubbles collision and rupture would make the hydrate formation rate tend to be constant. In the final period, the low subcooling degree (within 1.0 K) and low bubble strength would make bubbles possibly rupture. Afterwards, the sudden enlarge of the gas-liquid interface area would promote the hydrate growth rate sharply.

Notice that hydrate formation rate reduces significantly when XG solution concentration increase from 0% to 0.3%, and hydrate formation rate at XG solution concentration of 0.3% is 2–8 times lower than that of 0%, indicating XG solution inhibits the hydrate formation to some extent. Therefore, XG solution could be used as one kinetic inhibitor of the water-soluble polymer.

The change of hydrate formation rate with subcooling degree at different flow rates of XG solution is shown in Fig. 4.4b, showing the hydrate formation rate at 1.5 m/s is significantly higher than the corresponding rate at 1.18 m/s, because the larger flow rate would aggravate the gas-liquid mass transfer and facilitate the hydrates formation. It is also found that the hydrate formation rate decreases and increases rapidly at 5.0 K and 1.0 K respectively. Based on the two nodes, the areas from 6.0 K to 5.0 K and from 1.0 K to 0 K are marked as "start area" and "end area"; the area with constant hydrate formation rate of 5.0 K to 1.0 K is marked as "formation area". Ultimately, hydrate formation includes three stages: start area, formation area and end area. Such an area is caused by the combined action of hydrate formation inhibition from hydrate shell growth and hydrate formation facilitation from bubble breakup.

One adjustable parameter C_{HF}, named the hydrate formation integration constant, is added to the mass transfer model for experimental data match, and the hydrate formation rate could be expressed as Eq. 4.12. The parameter C_{HF} could be represented as the ratio of experimental value to the theoretical calculation value of the hydrate formation rate shown in Eq. 4.13. Specifically, for the hydrate formation inhibition from hydrate shell growth in the start area, the adjustable parameter $C_{HF,O}$ is proposed. Similarly, the adjustable parameter $C_{HF,F\text{-}E}$ is proposed to describe the influence of hydrate shell reconstruction and bubble rupture in the formation and end area. Therefore, the hydrate formation integration constant could be expressed as Eq. 4.14.

$$\frac{\mathrm{d}n_{CH_4}}{\mathrm{d}t} = \frac{\mathrm{d}n_H}{\mathrm{d}t} = C_{HF}\frac{1}{M_{CH_4}}k_{CH_4}aV_{tot}C_w(x_I - x_{eq}) \tag{4.12}$$

$$C_{HF} = \frac{(\mathrm{d}n/\mathrm{d}t)_{HF,exp}}{(\mathrm{d}n/\mathrm{d}t)_{HF,pred}} \tag{4.13}$$

$$C_{HF} = C_{HF,O} + C_{HF,F\text{-}E} \tag{4.14}$$

To reveal the relation between $C_{HF,O}$ and subcooling degree T_{sub}, the start time (6.0 K in the experiment) and the end time of hydrate formation is marked as 1.0 K and 0 K respectively.

The distribution of $C_{HF,O}$ with subcooling degree T_{sub} in different XG solution flow rates are shown in Fig. 4.5 [11], illustrating $C_{HF,O}$ decreases exponentially with the decrease of T_{sub}. This indicates the mass transfer rate between the gas-liquid phase would decrease due to hydrate shell formation on the bubble surface. The relation between $C_{HF,O}$ and T_{sub} could be fitted to the form of Eq. 4.15. Figure 4.6 illustrates that the linear relations between α, β and XG solution concentration could be expressed by Eqs. 4.16 and 4.17, and the linear tendency between a, b, c, d, and flow rate v_{XG} could be expressed by Eq. 4.18. Substituting Eq. 4.18 into Eqs. 4.16 and 4.17 could calculate the expressions of α and β. Ultimately, $C_{HF,O}$ can be expressed as Eq. 4.19.

$$C_{HF,O} = \alpha \exp(\beta T_{sub}) \tag{4.15}$$

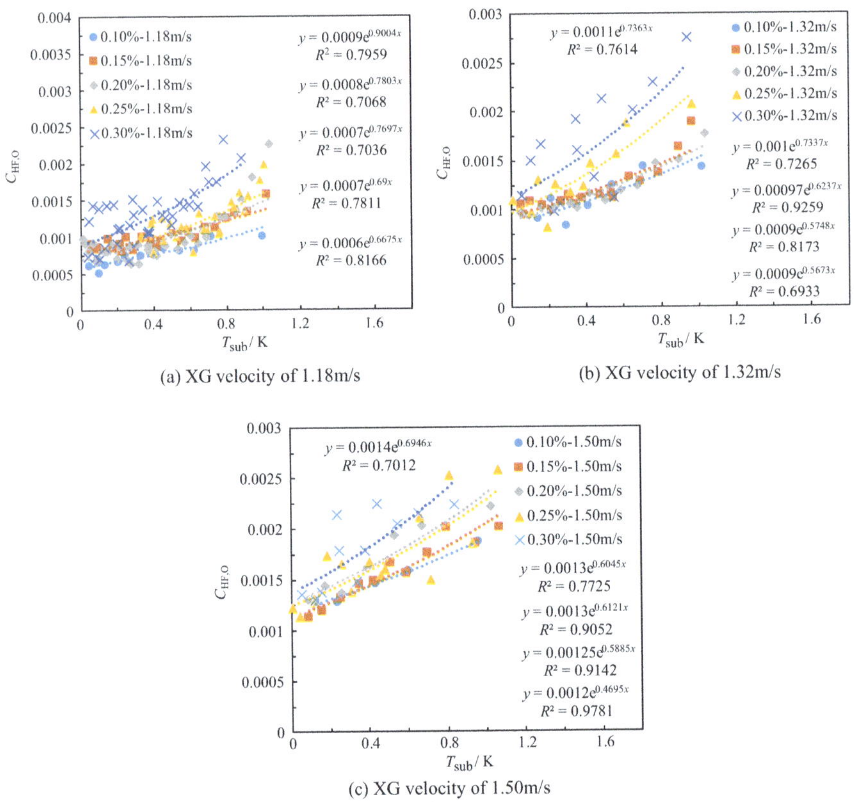

Fig. 4.5 Distribution of $C_{HF,O}$ with T_{sub}

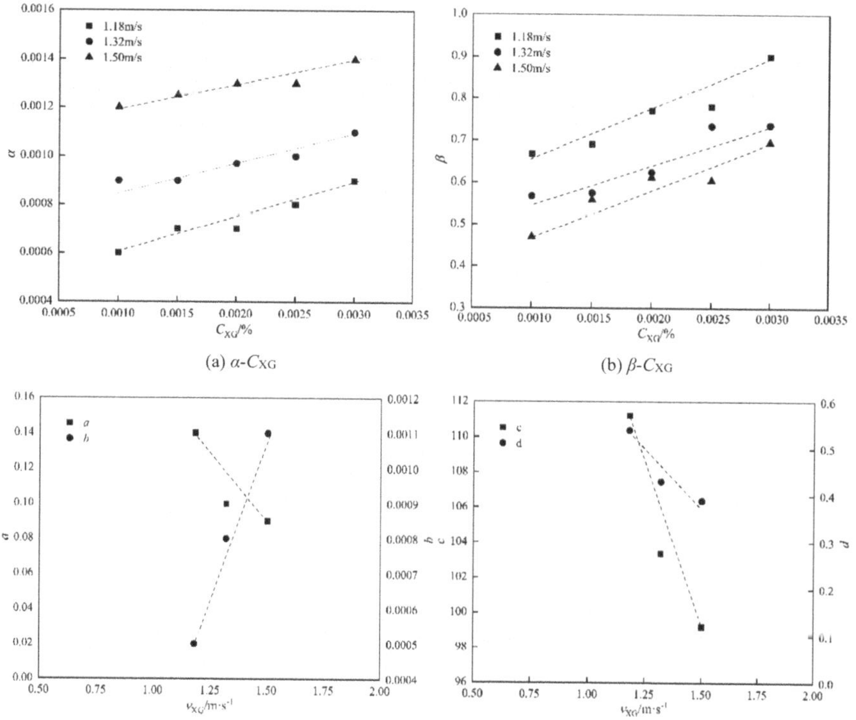

(a) α-C_{XG} (b) β-C_{XG}

(c) a, b- v_{XG} (d) c, d-v_{XG}

Fig. 4.6 Regression parameters α, β, a, b, c, d of $C_{HF,O}$

$$\alpha = aC_{XG} + b \tag{4.16}$$

$$\beta = cC_{XG} + d \tag{4.17}$$

$$\begin{cases} a = -0.1516v_{XG} + 0.3121 \\ b = 0.0019v_{XG} - 0.0017 \\ c = -36.764v_{XG} + 153.63 \\ d = -0.4566v_{XG} + 1.0618 \end{cases} \tag{4.18}$$

$$C_{HF,O} = [(-0.1516v_{XG} + 0.3121)C_{XG} + (0.0019v_{XG} - 0.0017)] \times e^{[(-36.764v_{XG}+153.63)C_{XG}+(-0.4566v_{XG}+1.0618)]T_{sub}} \tag{4.19}$$

where α is the coefficient of an adjustable parameter; β is the index of an adjustable parameter; a shows the coefficient of α; b shows the intercept of α; c shows the coefficient of β; d shows the intercept of β.

Likewise, the distribution of $C_{HF,F\text{-}E}$ with subcooling degree T_{sub} at different XG solution flow rates are shown in Fig. 4.7. From the figure, $C_{HF,F\text{-}E}$ increases with the hydrate formation, which indicates the bubbles breakup caused by hydrate shell modification and shear force could improve the hydrate formation rate. Moreover, $C_{HF,F\text{-}E}$ could be approximated as a power function of the subcooling degree T_{sub} shown in Eq. 4.20. Figure 4.8 illustrates that the linear relations between α, β and XG solution concentration could be expressed by Eqs. 4.21 and 4.22, and the linear trend between a, b, c, d, and flow rate v_{XG} could be expressed by Eq. 4.23. Substituting Eq. 4.23 into Eqs. 4.21 and 4.22 could calculate the expressions of α and β. Ultimately, $C_{HF,F\text{-}E}$ can be expressed as Eq. 4.24.

$$C_{HF,F\text{-}E} = \alpha T_{sub}^{\beta} \tag{4.20}$$

$$\alpha = aC_{XG} + b \tag{4.21}$$

$$\beta = cC_{XG} + d \tag{4.22}$$

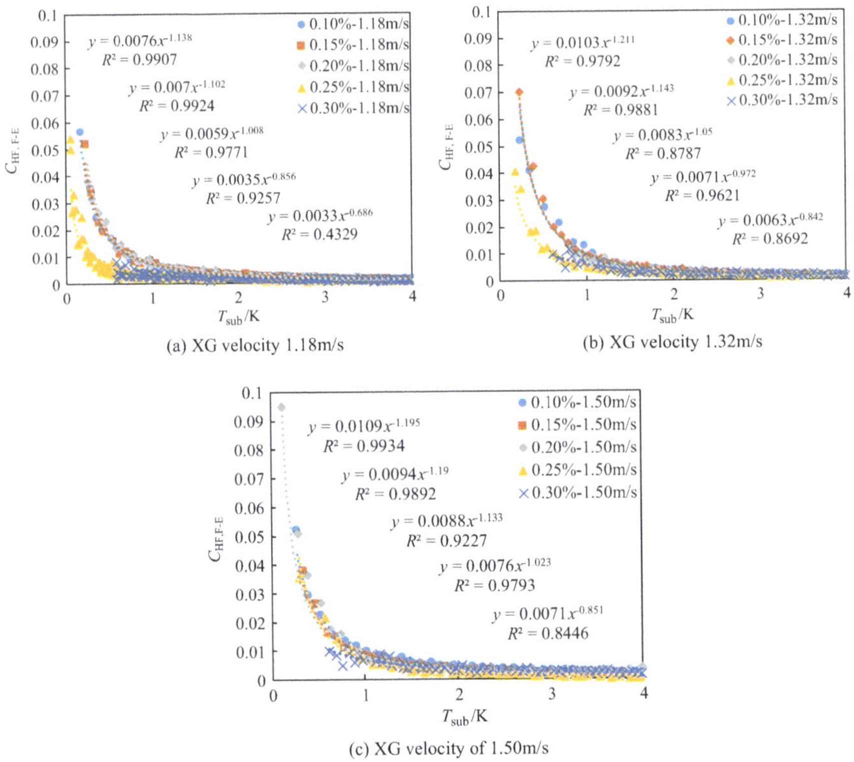

Fig. 4.7 Distribution of $C_{HF,F\text{-}E}$ with T_{sub}

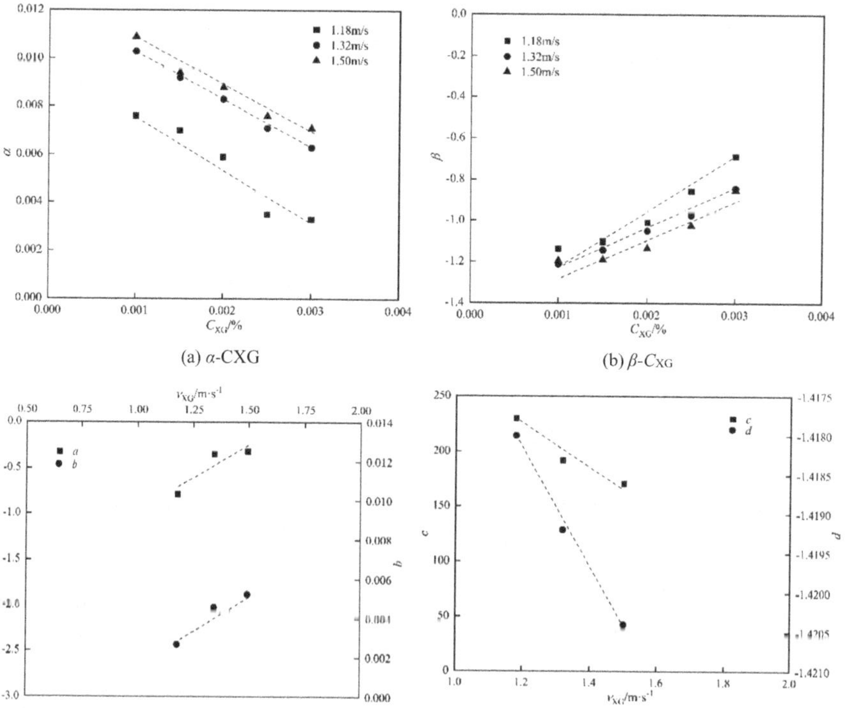

(a) α-CXG (b) β-C_{XG}

(c) a, b-v_{XG} (d) c, d-v_{XG}

Fig. 4.8 Regression parameters α, β, a, b, c, d of $C_{HF,F\text{-}E}$

$$\begin{cases} a = 1.6451 v_{XG} - 4.3001 \\ b = 0.0066 v_{XG} + 0.0029 \\ c = -181.17 v_{XG} + 439.15 \\ d = -0.0075 v_{XG} - 1.4093 \end{cases} \tag{4.23}$$

$$C_{HF,F\text{-}E} = [(1.6451 v_{XG} - 4.3001) C_{XG} + (0.0066 v_{XG} + 0.0029)] \times T_{sub}^{[(-181.17 v_{XG} + 439.15) C_{XG} + (-0.0075 v_{XG} - 1.4093)]} \tag{4.24}$$

4.2 Influence of Hydrate Phase Transition on Multiphase Flow During Deepwater Drilling

4.2.1 Well Annulus Multiphase Flow Model During Deepwater Drilling

During deepwater drilling, after gas invades into the well, the solid hydrate would form and then return upwards with drilling fluid, finally transform into gas and water when hydrate breaks away from the hydrate formation area. This phase transition process poses new problems and challenges to parameter design of drilling, production process and riser [23, 24].

Since the 1960s, many well annulus multiphase flow models have been established by scholars. In 1968, LeBlanc et al. [25] established the first mathematical model of well kick, which was a current-sharing model; in 1972, Records [23] proposed one well kick model considering the friction loss in well annulus; in 1981, Horberock et al. [24] established a flow model of well kick where the wellbore pressure was solved by momentum equation; in 1982, Santos [26] proposed a deepwater flow model during well kick based on gas-liquid slippage and friction loss in two-phase flow area, in which bubble flow was adopted and the flow friction loss is calculated by Orkiszewski method. In 1987, Nickens [27] obtained a multiphase flow model by analyzing the mud and gas equation of state, the mass and momentum variation of the gas-liquid phase. In 1991, Santos [28] applied the multiphase flow model to calculate the kick in a horizontal well and found the maximum allowable gas invasion after shut-in in a horizontal well was larger than that in a vertical well. In 1995, Ohara [29] proposed a deepwater well control model considering wellbore well annulus, gas layer, choke pipeline and gas-liquid phase flow, and divided the model into several sub-models for calculation. In 2002, based on the slug flow, supposing the downhole pressure maintained constant during the well kick, Nunes [30] established a well kick model and applied the iterative method to solve the pressure distribution and gas-liquid percentage in well annulus and choke pipeline per time step. In 2007, Gao et al. [31] considered the influence of hydrate decomposition and thermal change owing to the hydrate phase transition, then established the multiphase flow and heat transfer model of well during drilling. In 2009, Sun et al., Wang et al. [32] established the deepwater well kick model, and preliminarily discussed the influence of hydrate formation on the flow parameters during well kick; aiming at the hydrate phase transition in the wellbore, Wang et al. and Sun et al. [33–37] established the continuity equation, momentum equation and energy equation for each-phase fluid in well annulus during overflow.

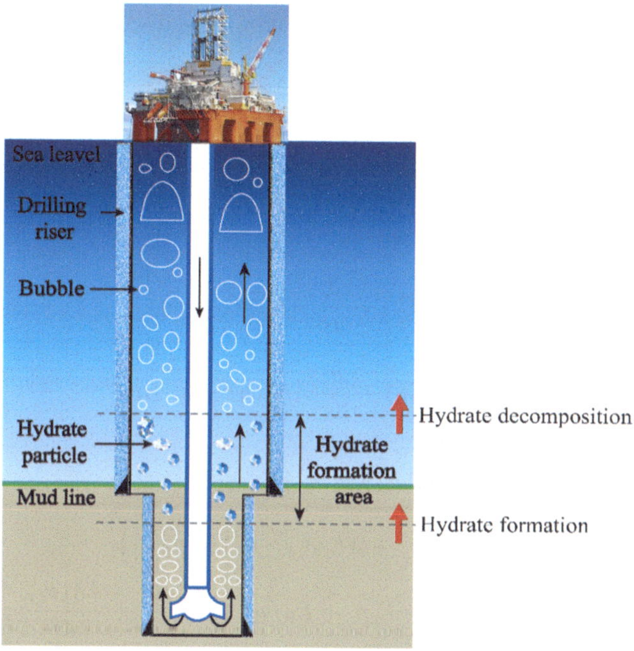

Fig. 4.9 Schematic diagram of multiphase flow in well annulus during deepwater drilling

4.2.1.1 Basic Assumptions

The schematic diagram of multiphase flow in well annulus during deepwater drilling is shown in Fig. 4.9, and the assumptions for the flow in well are as follows [35]:

(1) The each-phase fluid is composed of continuous particles;
(2) The physical properties difference between gas production from hydrate phase transition and stratum is neglected;
(3) The influence of chemical reagents on the hydrate phase transition is not considered during well kick;
(4) The thermal change caused by hydrate phase transition needs to be considered while establishing the energy equation;
(5) Gas, liquid and hydrate phases that play major roles in energy conversion are mainly considered when establishing the energy equation.

4.2.1.2 Equations of Multiphase Flow in Well Annulus

The equations governing the multiphase flow in well annulus mainly include the continuity equation, momentum conservation equation, and energy conservation equation of each component. The continuity equations of produced gas, liquid phase, cuttings, and hydrate are respectively shown in Eqs. 4.25–4.28 and the momentum equation is shown in Eq. 4.29; the energy equation of well section below mudline is shown in Eq. 4.30 [35, 37, 38].

$$\frac{\partial}{\partial t}\left(A\rho_{\mathrm{g}}E_{\mathrm{g}}\right)+\frac{\partial}{\partial s}\left(A\rho_{\mathrm{g}}v_{\mathrm{g}}E_{\mathrm{g}}\right)=q_{\mathrm{g}}-x_{\mathrm{g}}r_{\mathrm{H}} \tag{4.25}$$

$$\frac{\partial}{\partial t}(A\rho_{\mathrm{m}}E_{\mathrm{m}})+\frac{\partial}{\partial s}(A\rho_{\mathrm{m}}v_{\mathrm{m}}E_{\mathrm{m}})=-\left(1-x_{\mathrm{g}}\right)r_{\mathrm{H}} \tag{4.26}$$

$$\frac{\partial}{\partial t}(A\rho_{\mathrm{c}}E_{\mathrm{c}})+\frac{\partial}{\partial s}(A\rho_{\mathrm{c}}v_{\mathrm{c}}E_{\mathrm{c}})=q_{\mathrm{c}} \tag{4.27}$$

$$\frac{\partial}{\partial t}(A\rho_{\mathrm{H}}E_{\mathrm{H}})+\frac{\partial}{\partial s}(A\rho_{\mathrm{H}}v_{\mathrm{H}}E_{\mathrm{H}})=r_{\mathrm{H}} \tag{4.28}$$

$$\begin{aligned}&\frac{\partial}{\partial t}\left(AE_{\mathrm{g}}\rho_{\mathrm{g}}v_{\mathrm{g}}+AE_{\mathrm{m}}\rho_{\mathrm{m}}v_{\mathrm{m}}+AE_{\mathrm{c}}\rho_{\mathrm{c}}v_{\mathrm{c}}+AE_{\mathrm{H}}\rho_{\mathrm{H}}v_{\mathrm{H}}\right)\\&+\frac{\partial}{\partial s}\left(AE_{\mathrm{g}}\rho_{\mathrm{g}}v_{\mathrm{g}}^2+AE_{\mathrm{m}}\rho_{\mathrm{m}}v_{\mathrm{g}}^2+AE_{\mathrm{c}}\rho_{\mathrm{c}}v_{\mathrm{c}}^2+AE_{\mathrm{H}}\rho_{\mathrm{H}}v_{\mathrm{H}}^2\right)\\&+Ag\cos\theta\left(E_{\mathrm{g}}\rho_{\mathrm{g}}+E_{\mathrm{m}}\rho_{\mathrm{m}}+E_{\mathrm{c}}\rho_{\mathrm{c}}+E_{\mathrm{H}}\rho_{\mathrm{H}}\right)+\frac{\mathrm{d}(Ap)}{\mathrm{d}s}+\frac{\mathrm{d}A(F_{\mathrm{r}})}{\mathrm{d}s}=0\end{aligned} \tag{4.29}$$

$$\begin{aligned}&\frac{\partial}{\partial t}\left[\rho_{\mathrm{g}}E_{\mathrm{g}}\left(h+\frac{1}{2}v_{\mathrm{g}}^2-gs\cos\theta\right)\right]+\left[\rho_{\mathrm{m}}E_{\mathrm{m}}\left(h+\frac{1}{2}v_{\mathrm{m}}^2-gs\cos\theta\right)\right]A+\frac{r_{\mathrm{H}}\cdot A\Delta H_{\mathrm{H}}}{M_{\mathrm{H}}}\\&-\left\{\frac{\partial\left[W_{\mathrm{g}}\left(h+\frac{1}{2}v_{\mathrm{g}}^2-gs\cos\theta\right)\right]}{\partial s}+\frac{\partial\left[W_{\mathrm{m}}\left(h+\frac{1}{2}v_{\mathrm{m}}^2-gs\cos\theta\right)\right]}{\partial s}\right\}\\&=2\left[\frac{1}{A'}(T_{\mathrm{ei}}-T_{\mathrm{a}})-\frac{1}{B'}(T_{\mathrm{a}}-T_{\mathrm{t}})\right]\end{aligned} \tag{4.30}$$

where

$$\frac{1}{A'}=\frac{1}{2\pi}\frac{k_{\mathrm{e}}+r_{\mathrm{co}}U_{\mathrm{a}}T_{\mathrm{D}}}{r_{\mathrm{co}}U_{\mathrm{a}}k_{\mathrm{e}}} \tag{4.31}$$

$$\frac{1}{B'}=\frac{1}{2\pi r_{\mathrm{ti}}U_{\mathrm{t}}} \tag{4.32}$$

The energy equation of well section above mudline is the same as Eq. 4.30, but they have different expressions of A':

$$\frac{1}{A'}=\frac{1}{2\pi r_{\mathrm{co}}U_{\mathrm{a}}} \tag{4.33}$$

where E_{H}, E_{c}, E_{m}, E_{g} show the respective volume fraction of hydrates, cuttings, drilling fluids, and produced gas, %; v_{H}, v_{c}, v_{m}, v_{g} stand for the respective upward velocities of hydrates, cuttings, drilling fluids and produced gas in well annulus, m/s; ρ_H, ρ_c, ρ_m, ρ_g represent the respective density of hydrate, cuttings, drilling fluids, and produced gas, kg/m^3; q_g is the hydrate decomposition rate of unit length of the reservoir, kg/(s·m); q_c is the cuttings production rate per unit length, kg/s; x_g refers to the mass fraction of gas in hydrate, dimensionless; r_H stands for the hydrate formation/decomposition rate per length of the well annulus, kg/(s m); A is the cross-sectional area of well annulus, m^2; θ is the deviation angle of well, (°); F_r represents the friction in well annulus, Pa; p shows the pressure in well annulus, Pa; h refers to the enthalpy, including internal energy and kinetic energy, J; ΔH_H is the enthalpy of the gas hydrate phase transition, with its positive or negative value representing, J/mol; W_g, W_m show the mass flow rates of gas and fluid phases. kg/s; T_a, T_{ei}, T_t is the temperature in well annulus, stratum/seawater and drill string, °C; T_D is the function of transient heat transfer; r_{co} shows the outer diameter of the casing pipe, m; r_{ti} shows the initial inner diameter of well, m; U_a stands for the comprehensive heat transfer coefficient of annular fluid and hydrate formation, W/(m^2 K); U_t shows the comprehensive heat transfer coefficient of drill string, W/(m^2·K); k_e represents the thermal conductivity of hydrate formation, W/(m·K); s denotes the unit length, m.

4.2.1.3 Initial and Boundary Conditions

(1) Fixed Solution Conditions for Temperature Field

The fluid intrusion during drilling would change the former condition of fluid motion. Here, the temperature in well and drill string under steady-state could be the initial conditions of the instantaneous temperature field.

The fluid temperature at the inlet of drill string could be directly measured, so

$$T_c(0, t) = T_{in} \tag{4.34}$$

At bottom of well, the liquid temperature in drill string is the same as that in well annulus, so

$$T_c(H, t) = T_a(H, t) \tag{4.35}$$

The seawater temperature field is known and expressed as

$$T_s = T_0 + T_s(h_s, s) \tag{4.36}$$

The stratum temperature field is known and expressed as

$$T_G = T_s(H_s, s) + K_{Gr}h \tag{4.37}$$

where T_{in} shows the inlet temperature of drill string, °C; T_c is the temperature of the fluid in drilling string, °C; T_a shows the annular temperature, °C; T_s is the seawater temperature, °C; T_G is the temperature of hydrate formation, °C; K_{Gr} is the geothermal gradient, °C/m; H stands for the well bottom depth, m; H_s stands for the total depth of seawater, m; h refers to the well depth at a point, m; h_s refers to the seawater depth at a point, m.

(2) Fixed Solution Conditions for Pressure and Fluid Parameters

In case of normal drilling, when drilling to the top of the reservoir, there is no influx of stratum fluid:

$$E_H(S,0) = E_g(S,0) = E_c(S,0) = E_m(S,0) = 0 \tag{4.38}$$

$$E_c(S,0) = \frac{v_{sc}(S,0)}{C_c v_{mm}(S,0) + v_{cr}(S,0)} \tag{4.39}$$

$$E_m = 1 - E_c \tag{4.40}$$

$$v_m(S,0) = \frac{Q_m(S,0)}{A(s)E_m(S,0)} \tag{4.41}$$

$$v_c(S,0) = \frac{Q_c}{A(s)E_c(S,0)} \tag{4.42}$$

$$p(S,0) = p(S) \tag{4.43}$$

$$S = h + h_s \tag{4.44}$$

where v_{sc} is the apparent velocity of the cuttings, m/s; C_c is the velocity distribution factor; v_{mm} shows the velocity of the mixture, m/s; v_{cr} shows the slippage velocity of the cuttings, m/s; Q_m shows the drilling fluid circulation rate, m^3/s; Q_c refers to the volumetric flow rate of cuttings, m^3/s.

By solving the equations above, the initial multiphase flow parameters along the well such as pressure, velocity and each-phase volume fraction could be determined. Before killing well, the stratum pressure and weighted mud density shall be determined by measuring the stable vertical pressure, casing pressure and the volume increment of mud in mud pit after well shut-in. Besides, the each-phase velocity at one point is determined according to $V_{sl} = 0$ and slippage velocity. These parameters could be used as the initial fixed solution conditions during the well killing.

The overflow conditions during drilling are as below:

$$\begin{cases} p(0,t) = p_s \\ q_g(H,t) = q_g \\ q_c(H,t) = q_c \end{cases} \tag{4.45}$$

where q_c shows the volumetric flow rate of cuttings, m^3/s; q_g shows the gas production, m^3/d; P_s refers to the atmospheric pressure, Pa.

4.2.1.4 Auxiliary Equations

To enclose and solve the multiphase flow equations, it is necessary to establish related auxiliary equations for the multiphase flow process involving the hydrate phase transition. The auxiliary equations mainly comprise velocity equation, slippage velocity equation, fluid PVT equation, geometric equation, volume fraction equation, fluid viscosity equation, frictional loss equation, seawater temperature field equation, stratum temperature field equation, flow model and energy equation in the drill string. Overall, combining the aforementioned auxiliary equations and multiphase flow equations could make it possible to solve the simultaneous equations.

4.2.2 *Influence of Hydrate Phase Transition on Bubble Migration*

For one deepwater well, the seawater depth is 1500 m. Assume that the upward velocity of hydrate particle and one methane bubble with a diameter of 1 cm is 0.05 m/s and 0.2 m/s after stopping drilling. The hydrate formation area in the well after stopping drilling is shown in Fig. 4.10, illustrating the well temperature field in summer and winter are coincident approximately when well depth is greater than 200 m, and the hydrate formation area is 580–1500 m.

For the single methane bubble with a diameter of 1 cm, the hydrate phase transition curve is shown in Fig. 4.11 when it rises from 1500 m to wellhead, illustrating the hydrate formation rate is far greater than its decomposition rate in deepwater well. The hydrate is completely transformed by methane bubble at 4.2 min, and begin to decompose at point C until they end at point D (D′) with a respective duration of 155.9 min and 168.3 min in summer and winter. Besides, it is found that the depth of 91 m for complete hydrate decomposition in winter is smaller than that of 54 m in summer due to the lower seawater temperature and hydrate decomposition rate in winter.

For the single methane bubble with different diameters, the hydrate phase transition curve when it rises from 1500 m to wellhead is shown in Fig. 4.12, which shows that the bubble diameter influences the hydrate phase transition process in well and the gas-liquid interface area increases with bubble diameter, leading to the higher

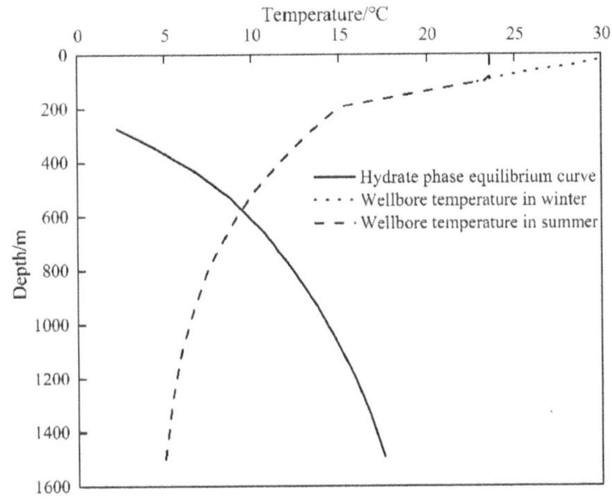

Fig. 4.10 Hydrate formation area in one wellbore after stopping drilling

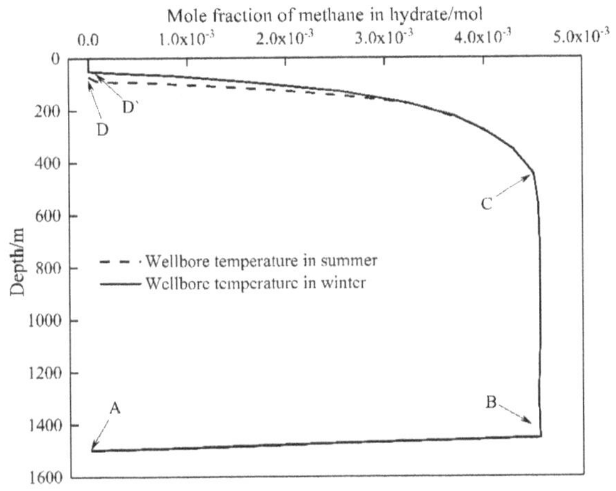

Fig. 4.11 Hydrate phase transition curve when the single methane bubble with diameter of 1 cm moves to wellhead

hydrate formation and decomposition rate. 2.4% of hydrate is still not decomposed when the methane bubble (diameter of 5 cm) moves to the wellhead, while all of the hydrates transforms into gas when it (diameter of 0.25 cm and 3 cm) moves to 183 m and 32 m respectively.

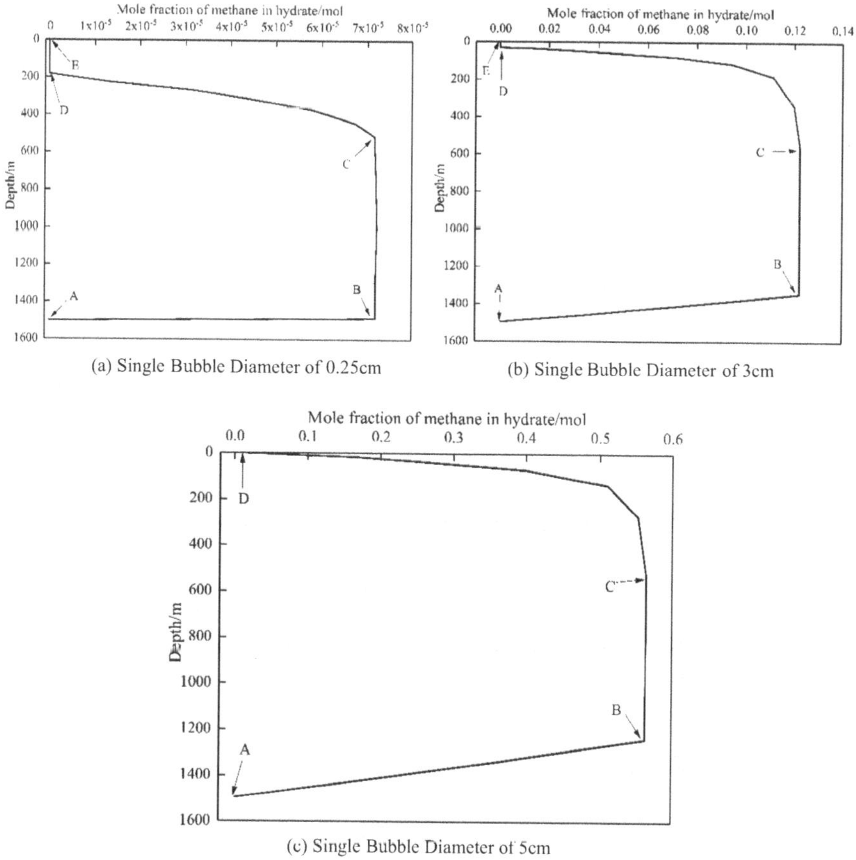

Fig. 4.12 Hydrate phase transition curve when the single methane bubble with different diameters moves to wellhead

4.2.3 *Influence of Hydrate Phase Transition on Well Multiphase Flow Without Inhibitors*

4.2.3.1 Influence of Hydrate Phase Transition on Gas Volume Fraction

When the overflow process lasts for 75 min, the changes of gas volume fraction with bubble diameter are shown in Fig. 4.13 [35], illustrating gas volume fraction increases gradually with bubble diameter during overflow, and hydrate phase transition has more and more effect on gas volume fraction with the decrease of average bubble diameter. This is because in certain gas volume fraction, the smaller bubble diameter means the greater gas-liquid interface area and hydrate formation rate. It is also found that there are inflection points of gas volume fraction curve at point A (near mudline) and point B, which represent the upper and lower boundary of the hydrate formation

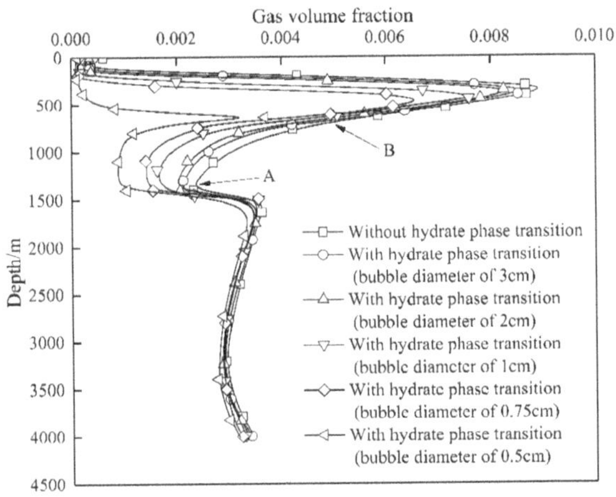

Fig. 4.13 Change of gas volume fraction with bubble diameter at overflow time of 75 min

area respectively. The cross-sectional area of vertical pipe above mudline (near point A) is greater than that of well annulus, then gas enters into the hydrate formation area. Both make the gas volume fraction decrease. Afterwards, the gas volume fraction begins to increase when gas gradually departs from the hydrate formation area.

Though Comprehensive analysis, we know that hydrate formation would reduce the gas volume fraction in well annulus and fluid level increment in mud pit, but also delay the overflow discovery. Therefore, "concealment" characteristics of gas intrusion and well kick during deepwater drilling is presented.

4.2.3.2 Influence of Hydrate Phase Transition on Fluid Level Increment in Mud Pit

The change of fluid level increment in the mud pit with overflow time at different bubble diameters is shown in Fig. 4.14 [35], showing the fluid level increment in mud pit would generally grow exponentially with overflow time. This is because as the gas enters in wellbore both pressure difference between the bottom hole and stratum, and the volume of well kick increase.

Notice that at the initial stage of overflow, the curves of fluid level increment in the mud pit with different bubble diameters overlap, then deviate from each other in 55 min and has a growing deviation degree, which is mainly caused by the large consumption of gas for hydrate formation. Furthermore, the smaller bubble diameter means the greater hydrate formation rate, and the more intensive effects of hydrate phase transition on fluid level increment in mud pit are found. In a word, the hydrate phase transition reduces the fluid level increment in the mud pit, which makes it difficult to detect the well kick early.

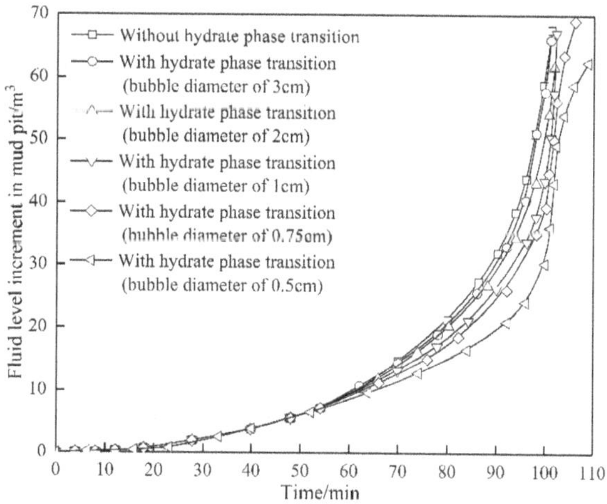

Fig. 4.14 Change of fluid level increment in the mud pit with overflow time at different bubble diameters

4.2.3.3 Influence of Hydrate Phase Transition on Bottom Hole Pressure and Shut-in Casing Pressure

The bottom hole pressure and shut-in casing pressure at different bubble diameters with overflow time are as shown in Figs. 4.15 and 4.16 [35], which show that hydrate phase transition delays the tendency of two pressure, and the smaller bubble diameter means more obvious delaying effect. Consequently, according to the difference of shut-in casing pressure and shut-in vertical pressure, it is are prone to misjudge the gas intrusion strength or intrusion type.

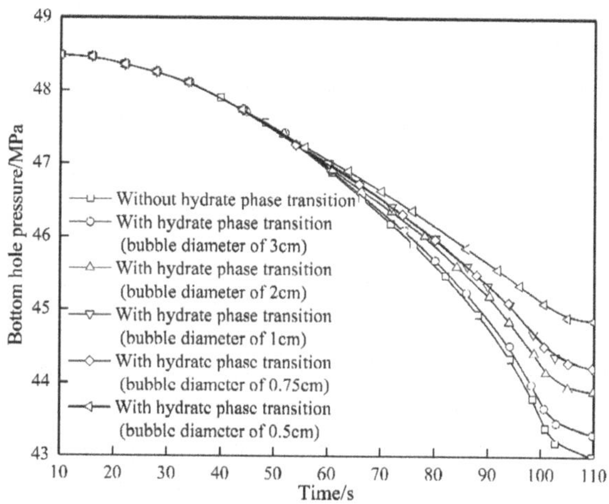

Fig. 4.15 Change of bottom hole pressure with overflow time at different bubble diameters

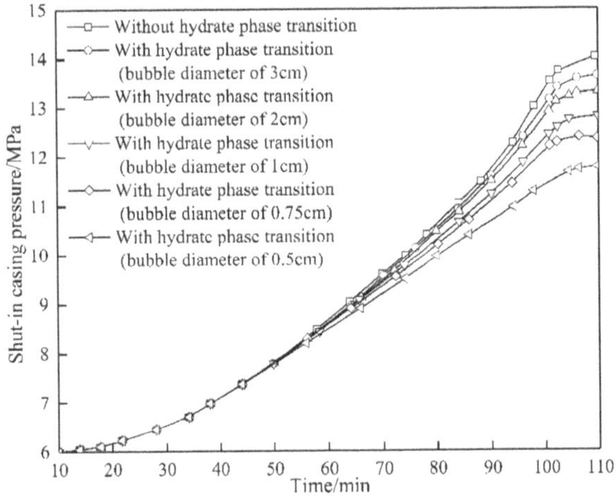

Fig. 4.16 Change of shut-in casing pressure with overflow time at different bubble diameters

4.2.3.4 Influence of Hydrate Phase Transition on Casing Pressure During Well Killing

Hydrate phase transition also exerts a certain influence on the killing process. For several initial conditions at the time of well shut-in, the fluid level increment in mud pit is 2 m^3 and the volumetric flow rate of well killing is 15 L/s. Besides, the driller's method is used to kill the well. The change of casing pressure with time during well killing is shown in Fig. 4.17, which shows that the casing pressure with considering the hydrate phase transition is lower than that without a hydrate phase transition. Hence, if the influence of hydrate phase transition is not taken into account during well killing, the casing pressure on-site will be greater than the actual value that we required, leading to the higher risks of stratum fracture.

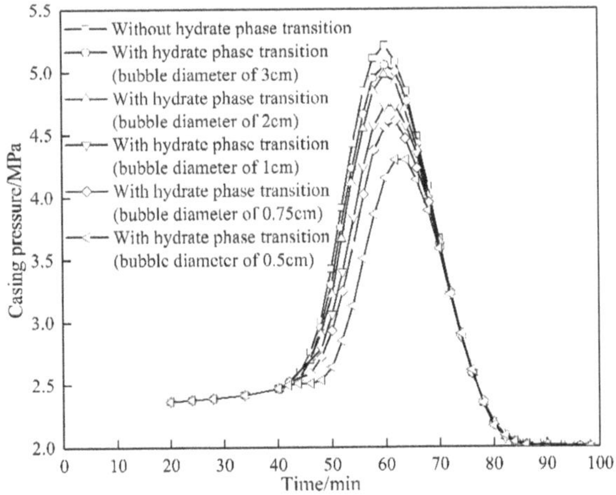

Fig. 4.17 Change of casing pressure with time during well killing

4.2.4 *Influence of Hydrate Phase Transition on Well Multiphase Flow with Inhibitors*

When injecting the methanol inhibitor, the influence of the hydrate phase transition on gas volume fraction and fluid level increment in mud pit are shown in Figs. 4.18 and 4.19 [35]. It indicates that both gas volume fraction and fluid level increment in mud pit increase with methanol concentration during overflow, but hydrate phase transition has less and less effect on gas volume fraction with the methanol concentration.

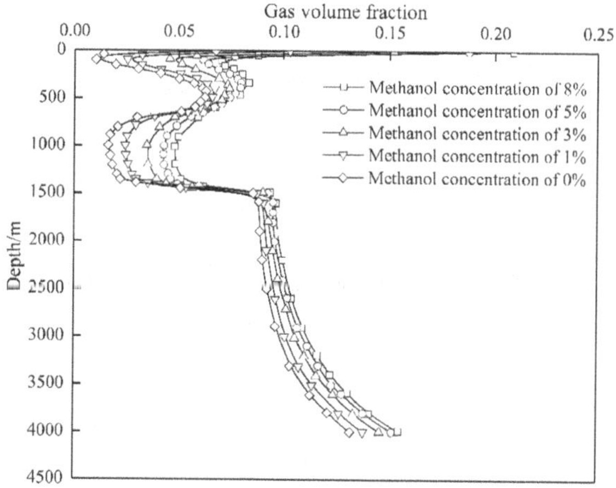

Fig. 4.18 Change of gas volume fraction with methanol concentration at the overflow time of 95 min

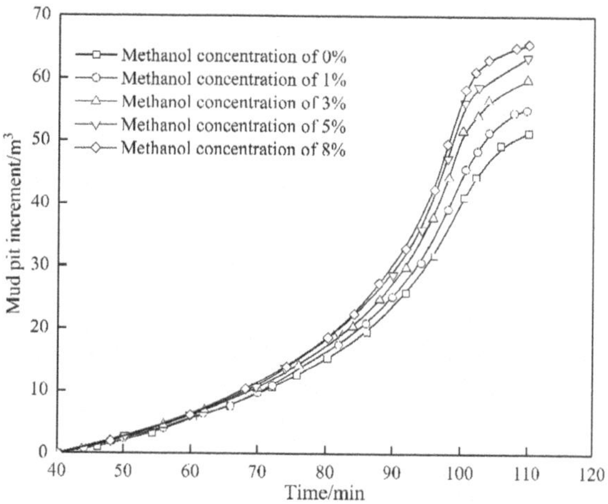

Fig. 4.19 Change of fluid level increment in mud pit with methanol concentration

References

1. Jing G, Shi B, Zhao J (2010) Natural gas hydrate shell model in gas-slurry pipeline flow. J Nat Gas Chem 19(3):261–266
2. Li XS, Yang B, Li G et al (2012) Experimental study on gas production from methane hydrate in porous media by Huff and Puff method in pilot-scale hydrate simulator. Fuel 94(1):486–494
3. Chen L, Sloan ED, Sum AK et al (2014) Methane hydrate formation and dissociation on suspended gas bubbles in water. J Chem Eng Data 59(4):261–283
4. Yang M, Zhe F, Zhao Y et al (2016) Effect of depressurization pressure on methane recovery from hydrate–gas–water bearing sediments. Fuel 166:419–426
5. Wang Z, Zhang J, Sun B et al (2017) A new hydrate deposition prediction model for gas-dominated systems with free water. Chem Eng Sci 163:145–154
6. Peysson Y, Nuland S, Maurel P et al (2003) Flow of hydrates dispersed in production lines. In: SPE annual technical conference and exhibition. Society of Petroleum Engineers
7. Peixinho J, Karanjkar PU, Lee JW et al (2010) Rheology of hydrate forming emulsions. Langmuir ACS J Surf Coll 26(14):11699–11704
8. Webb EB, Rensing PJ, Koh CA et al (2012) High pressure rheometer for in situ formation and characterization of methane hydrates. Rev Sci Instrum 83(1):015106
9. Webb EB, Koh CA, Liberatore MW (2013) Rheological properties of methane hydrate slurries formed from AOT + water + oil microemulsions. Langmuir 29(35):10997–11004
10. Fu W, Wang Z, Yue X et al (2019) Experimental study of methane hydrate formation in water-continuous flow loop. Energ Fuel 33(3):2176–2185
11. Fu W, Wang Z, Sun B et al (2019) Multiple controlling factors for methane hydrate formation in water-continuous system. Int J Heat Mass Transf 2018(131):757–771
12. Fu W, Wang Z, Duan W et al (2019) Characterizing methane hydrate formation in the non-Newtonian fluid flowing system. Fuel 253:474–487
13. Kawase Y, Hashiguchi N (1996) Gas-liquid mass transfer in external-loop airlift columns with newtonian and non-newtonian fluids. Chem Eng J Biochem Eng J 62(1):35–42
14. Deng Z, Wang T, Zhang N et al (2010) Gas holdup, bubble behavior and mass transfer in a 5 m high internal-loop airlift reactor with non-Newtonian fluid. Chem Eng J 160(2):729–737
15. Fu W, Wang Z, Sun B et al (2018) A mass transfer model for hydrate formation in bubbly flow considering bubble-bubble interactions and bubble-hydrate particle interactions. Int J Heat Mass Transf 127:611–621
16. Mohebbi V, Naderifar A, Behbahani RM et al (2012) Determination of Henry's law constant of light hydrocarbon gases at low temperatures. J Chem Thermodyn 51(10):8–11
17. Nakamura TMTST (2003) Stability boundaries of gas hydrates helped by methane-structure-H hydrates of methycychexane and cis-1, 2-dimethylcyclohexane. Chem Eng Sci 58:269–273
18. Kawase Y, Halard B, Moo-Young M (1987) Theoretical prediction of volumetric mass transfer coefficients in bubble columns for Newtonian and non-Newtonian fluids. Chem Eng Sci 42(7):1609–1617
19. Muthamizhi K, Kalaichelvi P, Powar ST et al (2014) Investigation and modeling of surface tension of power-law fluids. RSC Adv 4(19):9771–9776
20. Dodge DW, Metzner AB (2010) Turbulent flow of non-newtonian systems. AIChE J 5(2):189–204
21. Ma C, Chen G, Guo T (2002) Kinetics of hydrate formation using gas bubble suspended in water. Sci China, Ser B: Chem 45(2):208–215
22. Shi BH, Gong J, Sun CY et al (2011) An inward and outward natural gas hydrates growth shell model considering intrinsic kinetics, mass and heat transfer. Chem Eng J 171(3):1308–1316
23. Records LR (1972) Mud systems and well control. Petrol Eng 44(2):97–108
24. Hoberock LL, Stanbery SR (1981) Pressure dynamics in wells during gas kicks: Part 2—Component models and results. J Petrol Technol 33(8):1367–1378
25. Leblanc JL, Lewis RL (1968) A mathematical model of a gas kick. J Petrol Technol 20(8):888–898

26. Santos OA (1982) A mathematical model of a gas kick when drilling in deep waters. Colorado School of Mines
27. Nickens HV (1987) A dynamic computer model of a kicking well. SPE Drill Eng 2(2):159–173
28. Santos OLA (1991) A well operations in horizontal wells. SPE Drill Eng 6:111–117
29. Ohara S (1995) Improved method for selecting kick tolerance during deep water drilling operations. Louisiana State University, Baton Rouge
30. Nunes JOL, Bannwart AC, Ribeiro PR (2002) Mathematical modeling of gas kicks in deep water scenario. In: IADC/SPE asia pacific drilling technology. Society of Petroleum Engineers
31. Gao Y (2007) Study on multi-phase flow and well control of deepwater oil and gas drilling wellbore. China University of Petroleum
32. Wangzhiyuan S (2009) Multiphase flow behavior in well annulus with solid gas hydrate considering nature gas hydrate phase transition. Petrol Sci 6:57–63
33. Wang Z, Sun B, Gao Y et al (2008) Simulation calculation of well drilling by deepwater Driller's method. Acta Petrol Sinica 5:786–790
34. Wang Z, Sun B, Cheng H et al (2008) Prediction of natural gas hydrate formation area in deepwater drilling wellbore. Petrol Explor Dev 35(6):731–735
35. Wang Z, Sun B (2014) Deepwater gas kick simulation with consideration of the gas hydrate phase transition. J Hydrodyn 26(1):94–103
36. Sun B, Sun X, Wang Z et al (2017) Effects of phase transition on gas kick migration in deepwater horizontal drilling. J Nat Gas Sci Eng 46:710–729
37. Wang Z, Sun B, Gao Y et al (2010) Study on characteristics of well annulus multi-phase flow in hydrate reservoir drilling. J Basic Sci Eng 18(1):129–140
38. Wang Z, Sun B (2009) Annular multiphase flow behavior during deep water drilling and the effect of hydrate phase transition. Petrol Sci 6(1):57–63

Chapter 5
Mechanism and Prediction for Hydrate Deposition and Blockage in Deepwater Gas Well

During test and production in the deepwater gas well, the well flow mostly tends to be annular-mist flow with continuous gas phase and dispersed liquid phase. For the free water in the system, it is mainly in the form of liquid film on pipewall and droplet in the gas core, which could both form the solid hydrate at high pressure and low temperature. Moreover, the gas-liquid two-phase flow would be transformed into the gas-liquid-solid three-phase flow in well/pipeline owing to hydrate formation. As the continuous hydrates deposition, the gas-solid two-phase flow, and water-saturated gas single-phase flow appear [1, 2]. With hydrate formation and deposition, the flow patterns evolution in the vertical and horizontal pipeline are shown in Figs. 5.1 and 5.2 respectively.

The mechanisms for hydrate formation, deposition, and blockage vary considerably in different flow patterns, so in this chapter, based on the hydrate particles interaction, migration, and deposition, the corresponding model could be established to achieve the quantitative prediction for hydrate blockage in the deepwater gas well.

5.1 Hydrate Particles Interactions

For the mechanism for hydrate aggregation and deposition, it is very crucial to explore the hydrate particles' interaction at the micro-level. Considering van der Waals forces, capillary force and repulsive contact elastic force between hydrate particles, fluid-particle drag force based on Stokes's law, Yang et al. [3] simulated the hydrate particle aggregation in the pipe during flow with the two-dimensional discrete element model. The results indicate that there is no evident particle aggregation in the pipe when only considering van der Waals force, but the aggregate phenomena will be more obvious if the capillary force is considered additionally. Afterwards, Fidel-Dufour et al. [4] and Camargo et al. [5] believed when hydrate particles form a liquid bridge, the capillary force would generate and present as the main force for the irreversible

© Science Press and Springer Nature Singapore Pte Ltd. 2020

Z. Wang et al., *Natural Gas Hydrate Management in Deepwater Gas Well*,
https://doi.org/10.1007/978-981-15-6418-5_5

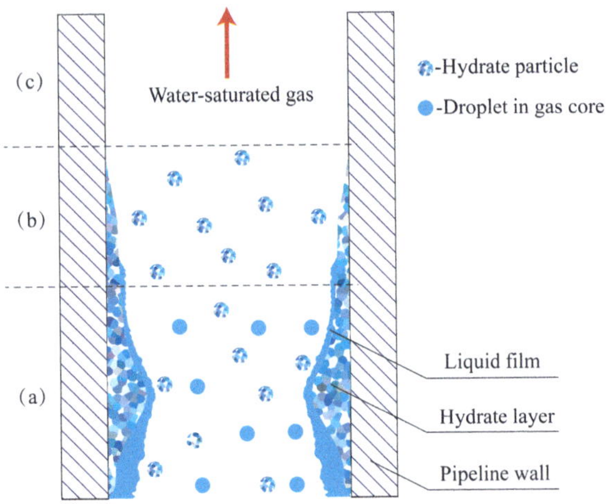

Fig. 5.1 Flow patterns evolution with hydrate formation and deposition in vertical pipeline, **a** gas-liquid-solid three-phase flow, **b** gas-solid two-phase flow, **c** water-saturated gas single-phase flow

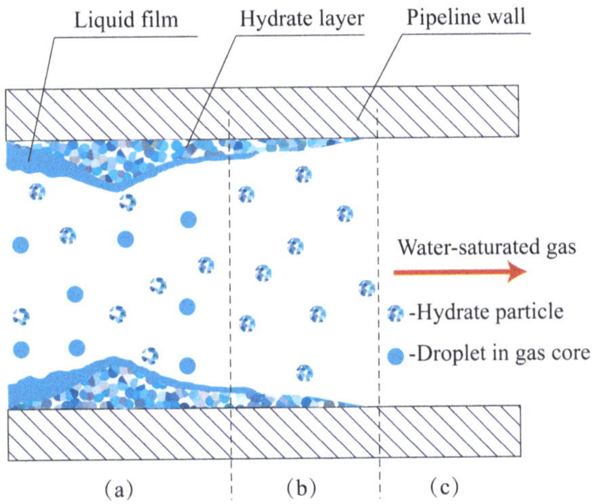

Fig. 5.2 Flow patterns evolution with hydrate formation and deposition in horizontal pipeline, **a** gas-liquid-solid three-phase flow, **b** gas-solid two-phase flow, **c** water-saturated gas single-phase flow

particle aggregation. However, free water and capillary force would not disappear until hydrate is completely formed, indicating the wettability of hydrate particles becomes the main factor for influencing the hydrate blockage. Zhu et al. [6] found the liquid bridge force increases with hydrate particle size and half-fill angle, and reduces with surface spacing when the surface spacing, half-fill angle and contact angle of the liquid bridge remain unchanged. However, the liquid bridge would not disappear until the surface spacing exceeded a certain distance. With one test instrument for micromechanical force, Yang et al. [3] firstly measured the adhesion force between ice-n-decane hydrate, ice-tetrahydrofuran hydrate and n-decane-tetrahydrofuran hydrate at different temperatures, which show the large approaching degree of cumulative distribution curves and obvious dispersity of adhesion force. By improving the Yang test device, Taylor et al. [7] studied the adhesion force between tetrahydrofuran hydrate particles in n-decane, and found the adhesion force is proportional to the contact force, contact time, fluid interface and temperature.

To stabilize the tetrahydrofuran hydrate particles, the adhesion force tests are conducted at a temperature below 0 °C, but ice generation could affect the experimental results. Therefore, cyclopentane (CyC5) is mostly selected as the hydrate former in subsequent studies. Dieker et al. [8] tested the capillary cohesion force of CyC5 hydrate particles in the presence of a little crude oil ($\leq 8\%$), and found crude oil could reduce their capillary cohesive force, but the adhesion force would increase if the surface-active substances of crude oil is removed. Aman et al. [9] tested the interaction force between CyC5 hydrate particles in mineral oil systems with surfactant and amphiphilic polymer (viscosity of 200cp), and found that the amphiphilic polymer and commercial naphthenic acid mixture could significantly suppress the gathering trend of hydrate particles. In 2011, Aman et al. [9] thought that the adhesion force is mainly dominated by capillary force when the contact time of hydrate particles is less than 30 s, or else mainly dominated by the tensile strength of hydrate. The tensile strength is about 0.91 MPa and the cohesive force of hydrate particles in the gas phase is about 9.1 mN/m at 3 °C. In 2012, Aman et al. [10] put forward that all the surfactants could be used to reduce the oil-water interfacial tension, while only some of the hydroxy acids could reduce the capillary cohesion force. Meanwhile, for completely generated hydrate particles, Aman et al. [11] pointed out the adhesion force in the gas phase is about 2 times and 6 times than that in liquid hydrocarbon and water. In the water phase, the adhesion form of hydrates is mainly composed of solid-solid absorption. Besides, the free water on hydrate particle surface would result in the fact that the cohesive force is increased by three times approximately, which further verify that liquid bridge has a significant influence on the capillary cohesion force. Based on this, Liu et al. [12] proposed an improved pendular liquid bridge model.

Maeda et al. [13] studied the adhesion hysteresis laws among CyC5 hydrate particles and hydrate micro-bulges, and noted that the temperature has no obvious effect on the adhesion hysteresis. Hydrate shell ruptures under great preload force, then the unconverted water flows out to form hydrate micro-bulges. With high-pressure test devices for micromechanical force, Lee et al. [14] measured the capillary cohesion force in the CH_4/C_2H_6 system and found the value is about 10 times that in the CyC5 system. Song et al. [15] proposed that in the cationic and anionic surfactant system, both initial contact force between hydrate particle-droplet and total adhesion energy decrease with the surfactant concentration. However, in the nonionic polymer inhibitors system, the adhesion energy and adhesion force do not change along with the inhibitor concentration. Cha et al. [16] studied the influence of droplet volume on the adhesion force of CyC5 hydrate particles-droplet, and proposed that their contact force constantly increases with droplet volume when droplet volume is less than 300 μl, then it turns to increase slowly. Liu et al. [17] clarified that the liquid bridge force of hydrate particles comprise the static liquid bridge force and dynamic viscosity force, and noted that liquid bridge force could be reduced and even eliminated by increasing the contact angle of water with hydrate particle, half-fill angle, the surface spacing of particles and reducing the particles size. Thus, to reduce the liquid bridge force and prevent the hydrate particles aggregation, corresponding measures of adding surfactant (to increase the contact angle of water with hydrate particle) and anti-agglomerant (to control particles size), increasing the water content in the system (to increase the half-fill angle) should be taken.

The studies above show that during hydrate formation, the interaction of hydrate particle-particle, hydrate particle-droplet, and hydrate particle-droplet-hydrate particle have significant impacts on the aggregation and deposition of dispersion phase, so measuring the interaction force could be regarded as one way to describe the related behavior and trend for hydrate particles aggregation [18]. Based on this, Liu et al. [19, 20] proposed a new method for directly measuring the interaction between CyC5 hydrate particles-water bridges with a micromechanical force test instrument, which is mainly introduced in this section.

5.1.1 Interaction Force Between Hydrate Particles

5.1.1.1 Experimental Process

The type II of hydrate mainly formed from CyC5 and water is the same as the actual hydrate in the pipeline. Furthermore, the solubility of CyC5 in water is similar to that of natural gas such as methane and ethane. However, the interaction force measurement experiments of CyC5 hydrate is allowed to be carried out at normal pressure, which has low requirements for equipment with simple and safe procedures. Therefore, CyC5 hydrate is selected as the experimental object.

In interaction force measurement experiments of CyC5 hydrate, glass fiber tends to penetrate the entire droplet. Though droplet is fully converted into hydrate particle, an exposed tip is left at their contact end still, which stops two particles from directly measuring the interaction force using glass fiber arranged horizontally (simply as "horizontal measurement"), but only allows to measure the interaction force using glass fiber arranged vertically (simply as "vertical measurement").

The schematic of the interaction force measurement between hydrate particles using "vertical measurement" is shown in Fig. 5.3. It is assumed that only the interaction force of F exists along the spherical center of particles, and its component of F in the vertical direction is the measured force.

$$F_{\mathrm{y}} = k\delta = F\sin\theta \tag{5.1}$$

To prevent the unwanted impact of the exposed tip-end of glass fiber and achieve the direct measurement of interactions between hydrate particles, the location of

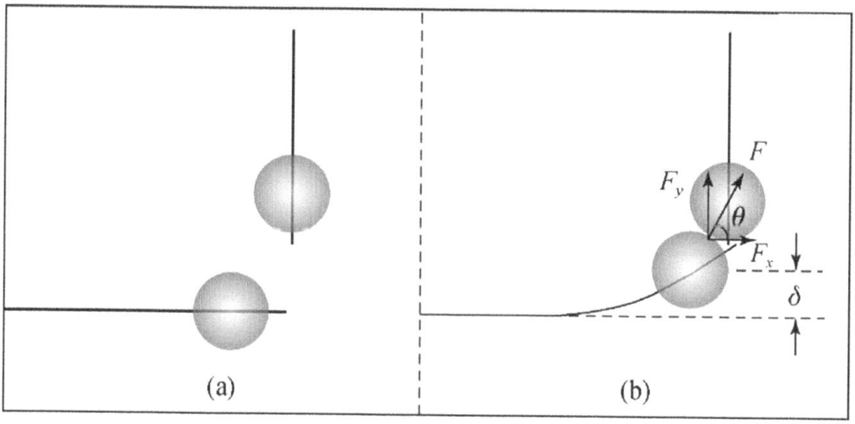

Fig. 5.3 Schematic of interaction force measurement between hydrate particles using "Vertical Measurement"

two glass fiber needs to be changed from the perpendicular to the parallel. That is, the interaction force between hydrate particles by "horizontal measurement" (see Fig. 5.4) enables them to be carried out. Schematic of hydrate particle formation using ice particle induction is shown in Fig. 5.5. Firstly, a dropper is used to place a droplet (on the order of 10^2 μm) on the end of the glass fiber cantilever, and the droplet is quenched in liquid nitrogen for the 20 s to completely generate ice particle. Afterwards, the ice particle is placed immediately in the cooling CyC5 bath. Finally, the ice particle begins to melt and the hydrate particle starts to form.

After the equilibration period lasts for 30 min, the hydrate-hydrate interaction measurement is performed and the hydrate particles appear to be completely rigid.

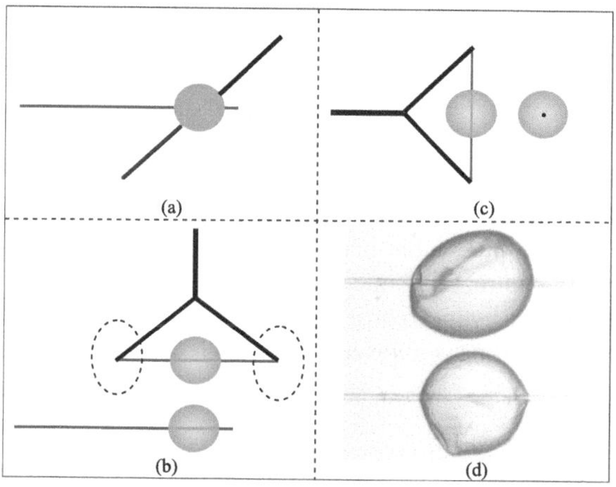

Fig. 5.4 Schematic of interaction force measurement between hydrate particles using "Horizontal Measurement"

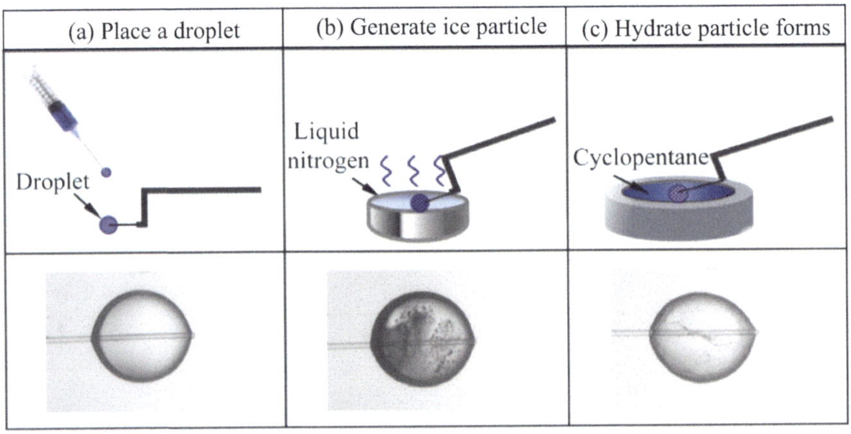

Fig. 5.5 Schematic of hydrate particle formation using ice particle induction

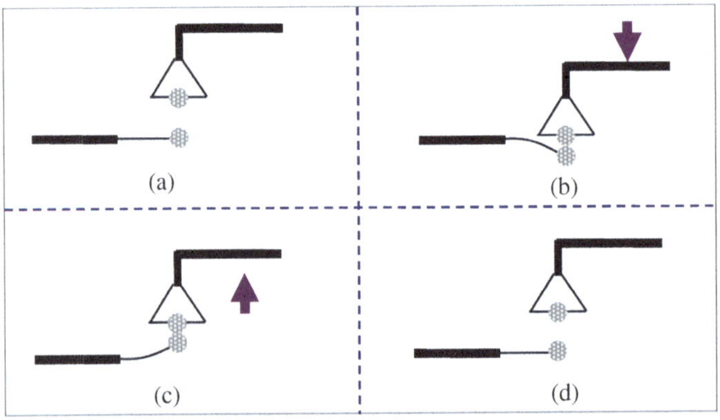

Fig. 5.6 Sequence of steps in hydrate–hydrate interaction measurements

The sequence of steps in the measurement is shown in Fig. 5.6, which includes "approaching" and "retracting" periods. During the approaching period, one hydrate particle is slowly brought into contact with the other one, creating a preload force on the hydrate particle due to the bending of the left-hand glass fiber cantilever (Fig. 5.6a, b). While "retracting", the hydrate particle moves away from the other one slowly until both two hydrate particles is detached (Fig. 5.2c, d). Unless specified otherwise, no residence time is allowed before the retracting period.

5.1.1.2 Results Analysis

The measurement results for interaction force between hydrate particles at 3 °C are shown in Fig. 5.7, showing the capillary cohesion force using "horizontal measure-

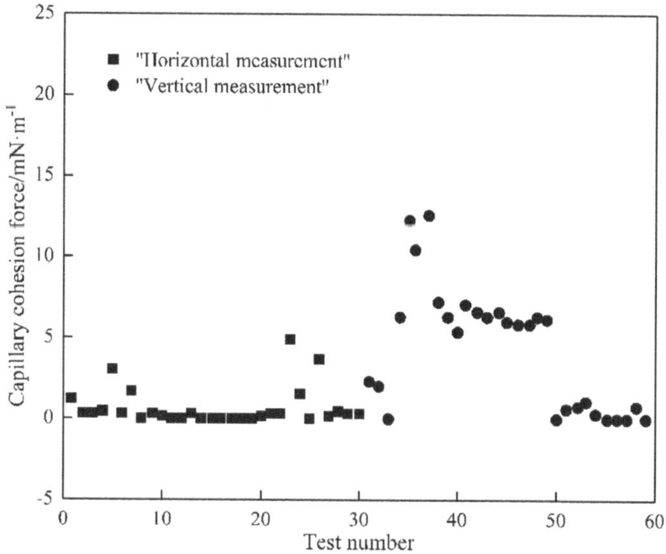

Fig. 5.7 Interaction force between hydrate particles at 13 °C

ment" is very small even close to zero. Compared to the "horizontal measurement", the cohesive force using "vertical measurement" is much larger generally, but there will be a situation where the capillary cohesion force is zero.

It has been reported that the cohesive force between hydrate particles is derived from the capillary force, and that is why the cohesive force between hydrate particles could be called the capillary cohesion force. That being said, there is one thin liquid film with dozens of nanometers on the hydrate particle. When one hydrate particle is brought into contact with the other one, liquid bridge form under the action of capillary force derived from the thin liquid film, but it is impossible to directly prove its existence.

By observing the hydrate particle image, a very rough growth surface of hydrate particles is captured and its roughness is much larger than the liquid film thickness, so the surface roughness of hydrate particles cannot be neglected for analyzing the interaction force between hydrates particles.

Schematic of "horizontal measurement" and "vertical measurement" considering the surface roughness of the hydrate particle is shown in Fig. 5.8. When performing "vertical measurement", a static friction force produced on the tangent of two particles could hinder its mutual movement, and component forces of the friction force along the direction of motion would result in the actual measured interaction force that is the resultant force of capillary force and frictional force. For "horizontal measurement", the static friction force is vertical to the motion direction and its component never exist at motion direction, so the actual measured interaction force mainly includes capillary force, which results in the fact that the interaction force by "horizontal measurement" is smaller than that by "vertical measurement".

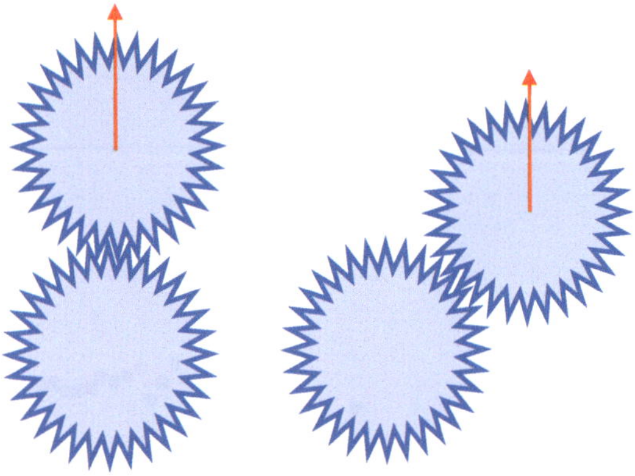

Fig. 5.8 Diagram of "horizontal measurement" and "vertical measurement" considering the surface roughness of hydrate particle

Notice that one large interaction force between hydrate particles using "horizontal measurement" could also be observed. One of the reasons will probably be that once a hydrate particle is brought into contact with the other one, its small protrusions on particle surface may be embedded in non-rigid shells. In other words, the surface roughness of the hydrate particle would generate an additional interaction force.

The capillary cohesive force between particles using "vertical measurement" at 3 °C is shown in Fig. 5.9, illustrating the range of capillary cohesive force at 3 °C is 2–5 mN/m, which is close to 4.3 mN/m described in works of literature. So far, what the capillary cohesive force derives from is still not explained appropriately. However, we will be more motivated to consider the surface roughness of hydrate particles and better explain such interaction force.

5.1.2 Interaction Force of Hydrate Particle-Droplet-Hydrate Particle

5.1.2.1 Experiment Process

To directly measure the hydrate particle-droplet-hydrate particle interactions, droplets have to be placed on a flat plate, but the characteristics of the plate (e.g., roughness) would affect the experimental results. Furthermore, after a hydrate particle is detached from one droplet, some water of the droplet will be detained on the particle surface to form the hydrate particle-droplet dimer. Herein, to eliminate the interference of the plate, on the basis of previous observations, the following procedures by forming another hydrate particle on the plate is proposed for directly measuring the hydrate particle-droplet-particle interactions.

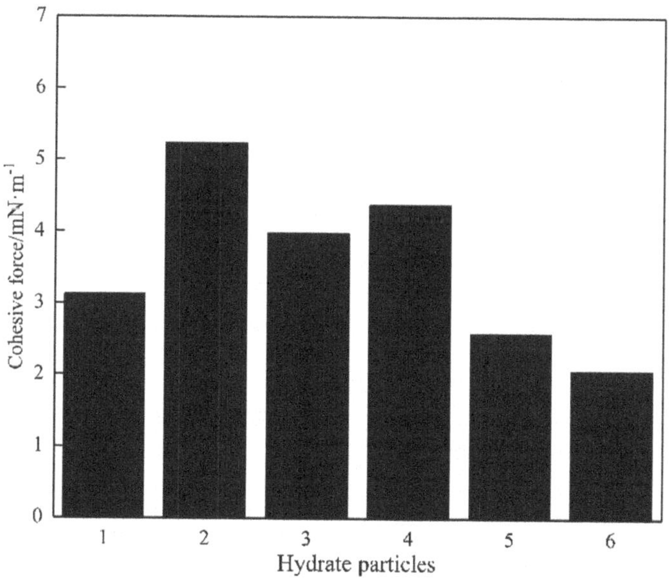

Fig. 5.9 Cohesive force between hydrate particles using "vertical measurement" at 3 °C

The entire measurement process included four successive steps (see Fig. 5.10):

(1) Generate two hydrate particles (Fig. 5.10a–c)
Two droplets are formed on the end of the glass fiber cantilever and one of the droplets is then transferred from one glass fiber to the Epoxy-A-coated layer in the air (Fig. 5.10a). Afterwards, the two droplets are quenched in liquid nitrogen for the 20 s to completely convert them into ice particles (Fig. 5.10b). The ice particles are placed immediately into the CyC5 bath within 2 s. The liquid CyC5 temperature is then slowly increased above the ice freezing temperature (0 °C) with a heating rate of 0.1 °C/min, allowing the ice particles to convert into CyC5 hydrate particles (Fig. 5.10c), which are immersed in the CyC5 cell at experimental temperatures for 30 min.

(2) Mount a droplet on the plate
Another water drop is formed on the end of the spare glass fiber cantilever, then it is transferred to the Epoxy-A-coated plate (Fig. 5.10d).

(3) Generate a hydrate-droplet dimer
The plate position (Fig. 5.10e) is adjusted while contacting the droplet and hydrate particle on the stationary glass fiber (Fig. 5.10f). The plate is then retracted quickly. Hence, some of the water is left on the hydrate particle surface, forming a hydrate particle-droplet dimer (Fig. 5.10g).

(4) Measure the hydrate particle-droplet-hydrate interactions
Adjust the plate position quickly (Fig. 5.10h), and then begin the measurement. The measurement procedure included "approaching" and "retracting". During the approaching process (Fig. 5.10h, i), the hydrate particle on the plate is slowly

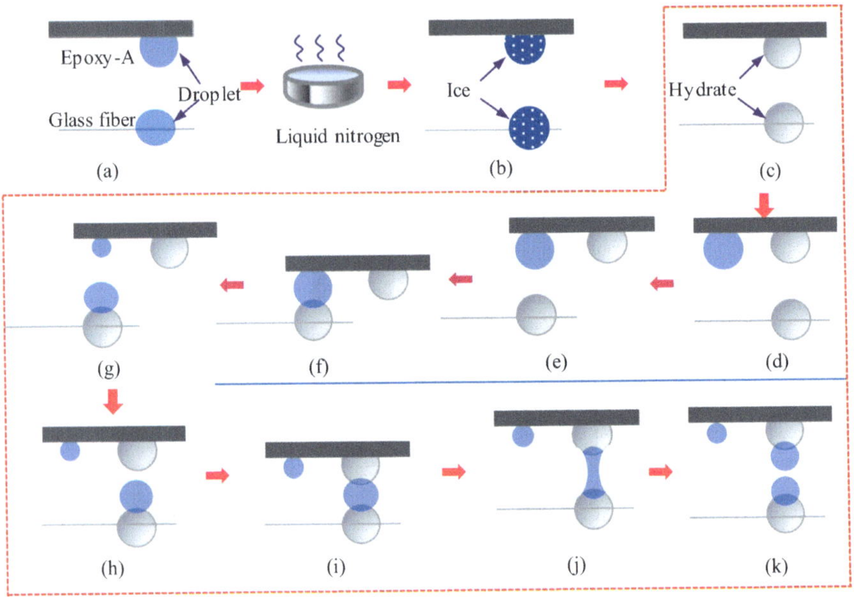

Fig. 5.10 Schematic of the proposed experimental procedure to directly measure the hydrate particle-droplet-hydrate particle interactions. The procedures in the red dotted lines indicate that the measurements are conducted in liquid CyC5

brought into contact with the hydrate particle-droplet dimer on the glass fiber, creating a preload force. During the retracting process, the hydrate particle on the plate moves away from the other hydrate particle slowly until the hydrate particle-droplet-hydrate particle complex detaches (Fig. 5.10j, k). The displacement of glass fiber cantilever during the measurement is captured at 30 frames per second in real-time. The displacement is multiplied by the cantilever spring constant to determine its interaction.

5.1.2.2 Results Analysis

Typical interaction force curve of the CyC5 hydrate particle-droplet-hydrate particle interaction and corresponding microscope images at various key locations in pure CyC5 at 6 °C are shown in Fig. 5.11, which shows the interaction force of hydrate particle-droplet-hydrate particle as a function of the plate displacement. Before the interaction, the plate is adjusted to set the positions of hydrate particle and hydrate-droplet dimer, as shown in point A. Then the plate slowly approaches the hydrate particle-droplet dimer, and no measurable net interaction force acts upon the two surfaces until point B, in which the droplet rapidly spreads on the top hydrate particle surface. Afterwards, a hydrate-CyC5.water three-phase-contact (TPC) line is formed and pinned on the top hydrate particle surface, resulting in a rapid drop in the interaction force from point B to C (initial contact force), which is ascribed to the hydrophilic

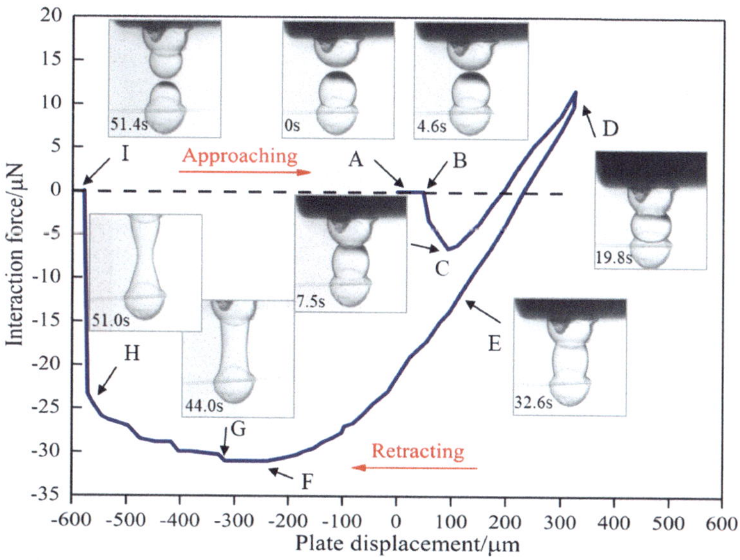

Fig. 5.11 Typical interaction force curve of the CyC5 hydrate particle-droplet-hydrate particle interaction

properties of the hydrate particle. As shown in microscope image of point C, the original droplet then behaves as a liquid bridge that bonds the two hydrate particles. As the plate continued to approach the bottom hydrate particle, the liquid bridge gradually changes from a stretching to a compression state. Correspondingly, the interaction force between the particles shifts from an attractive force to a repulsive force. At point D, the repulsive force reaches the set preload force and the plate starts to retract away from the bottom hydrate particle. As the distance between particles increases reversely, the interaction force first presents an approximately straight line (point D to E), then it curves from point E to F owing to the large liquid bridge deformation. At point F, the interaction force reaches the maximum level of adhesion force and remains constant until point G. Afterwards, the interaction force begins to decrease and the bottleneck appears in the liquid bridge as the liquid bridge is stretched further. Finally, a significantly narrow neck is formed and ruptured at point H. The liquid bridge is distributed between the two hydrate particles, forming two hydrate particle-droplet dimmers, and the shape of droplets is approximated as a perfect sphere.

During the entire measurement, both TPC lines are pinned on the hydrate particle surface due to strong wetting hysteresis between CyC5 hydrate and water. Furthermore, the outlines of liquid bridge intersecting with the hydrate particles presents a smooth curve, suggesting that hydrate is not visibly converted from the liquid bridge under this condition.

The same approaching-retracting procedure is repeated three times for the two hydrate-droplet dimers, and their results are shown in Fig. 5.12. The maximum

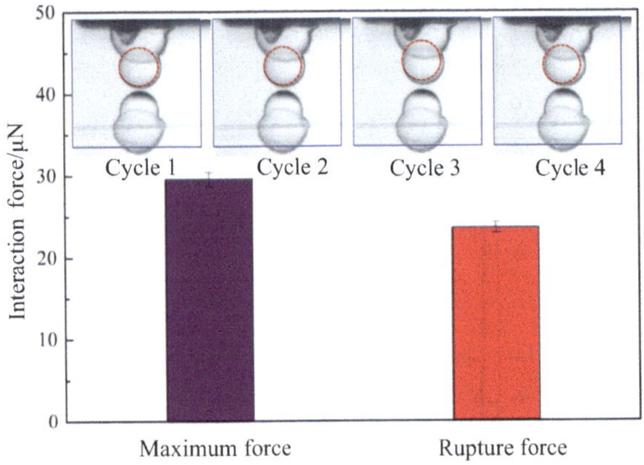

Fig. 5.12 Reproducibility measurements of interaction forces between hydrate particle-droplet-particle at 6 °C

interaction force and rupture force exhibits good repeatability, and the variation coefficients of the forces are less than 3%. The corresponding images also suggest that the amount of liquid distributed to two hydrate particles is approximately the same for the four test cycles. The high repeatability confirmes the reliability of the experimental apparatus and measurement procedure. Meanwhile, the repeatability further implies that the total liquid amount remains approximately constant during the entire measurement at 6 °C, that is, no visible hydrate is converted from the liquid bridge. Besides, the measured adhesion force between hydrate particles has obvious fluctuations. As suggested by Yang et al. [3, 21], the fluctuations arise because the nanoscale surface asperities have a slightly different alignment during each repeated contact, and a given surface roughness distribution therefore leads to a probabilistic distribution of the measured forces. In addition, the higher temperature, the greater the adhesion force is.

The temperature/subcooling degree plays an important role in hydrate-droplet interactions. Herein, the interaction behavior of CyC5 hydrate particle-droplet-hydrate particle and corresponding microscope images at various key locations in pure CyC5 at a lower temperature of 1.5 °C is shown in Fig. 5.13. During the preparation process of hydrate particle-droplet dimer, although the proceeding time is reduced as much as possible, a significant amount of water is still converted into hydrate due to the high subcooling degree and induction of the bottom hydrate particle (point A). At point B, the droplet contacts the top hydrate particle and subsequently spreads on the particle surface. However, because of the high subcooling level, the water molecules around the TPC line tends to rapidly form additional hydrates that prevent the effective wetting of water on the top hydrate particle. At point C, the contact area between the droplet and top hydrate particle is quite small despite the relatively large droplet volume. In the subsequent measurements, the hydrate continues to grow from both boundaries toward the middle. At point F, a significantly raised hydrate platform evolves around the contact region of the liquid bridge and top hydrate particle. Finally, the liquid bridge is broke, and a small amount of water is distributed to the top hydrate particle due to its relatively small contact area.

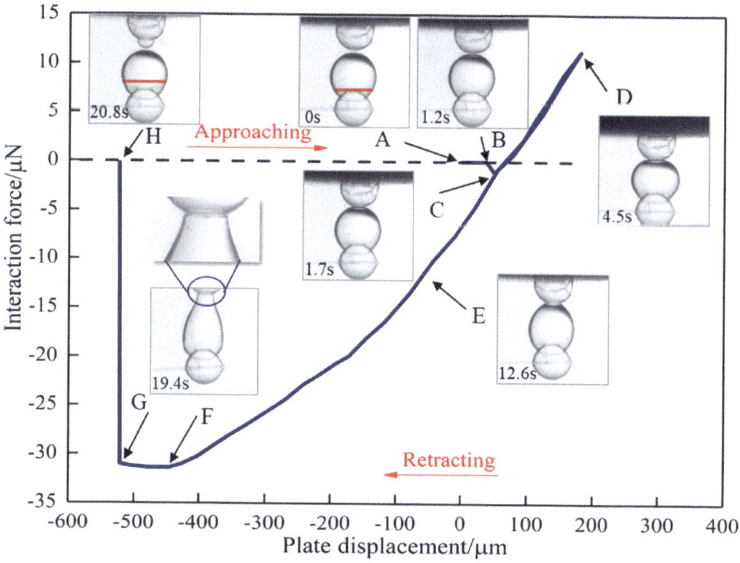

Fig. 5.13 Typical interaction force curve of the CyC5 hydrate particle-droplet-hydrate particle interaction and corresponding microscope images at various key locations in pure CyC5 at 1.5 °C. The red lines in the images denote the TPC line

Meanwhile, a large part of the droplet left on the bottom hydrate particle is encased by the hydrate shell.

In the water-in-oil dispersion system, the droplets sizes are not uniform, and they usually have a certain form of polydispersity. The state before and after the formation of hydrate particle-droplet dimer (before and after droplet rupture) is shown in Fig. 5.14. The liquid volume of droplets (spherical cap) attached to hydrate particles could be calculated by subtracting the volume of the spherical cap on the plate in Fig. 5.14b from that in Fig. 5.14a. The volume of the spherical cap is calculated as shown in Eq. 5.2. Where R shows the apparent radius of the droplet.

$$V_{\mathrm{d}} = \frac{4}{3}\pi R^3 - \frac{\pi}{6}\mathrm{h}\left(6Rh - 2h^2\right) \tag{5.2}$$

The effects of the water volume on the contact area and the normalized adhesion force of hydrate particle-droplet in pure CyC5 at 1.5 and 6 °C are shown in Fig. 5.15. For the measurements at 6 °C, at a relatively lower $V_{\mathrm{d}}/V_{\mathrm{p}}$, the increase in the contact area and adhesion forces with water volume is very slow, then both the contact area and adhesion forces increase dramatically with water volume. However, the contact areas at 1.5 °C are quite scattered. Herein, there was no clear relationship between the contact area and water volume. Furthermore, the contact areas at 1.5 °C are generally significantly smaller than those at 6 °C. Both phenomena are attributed to the higher rate of hydrate formation at 1.5 °C. As aforementioned, once the droplet contacts

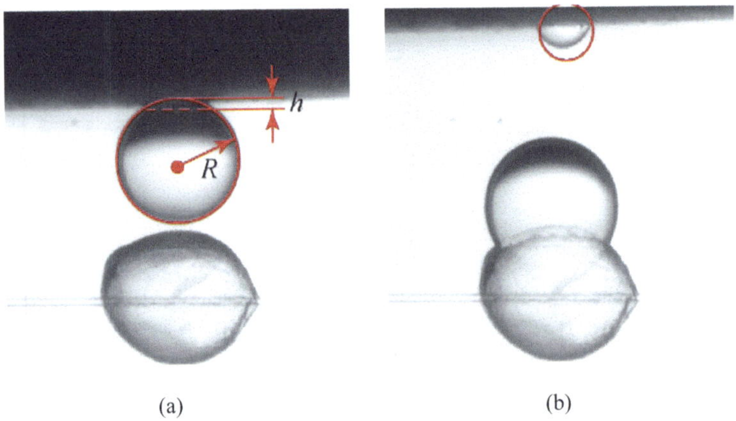

Fig. 5.14 The images before and after droplet rupture, **a** before droplet contacts the hydrate particle, **b** after droplet rupture

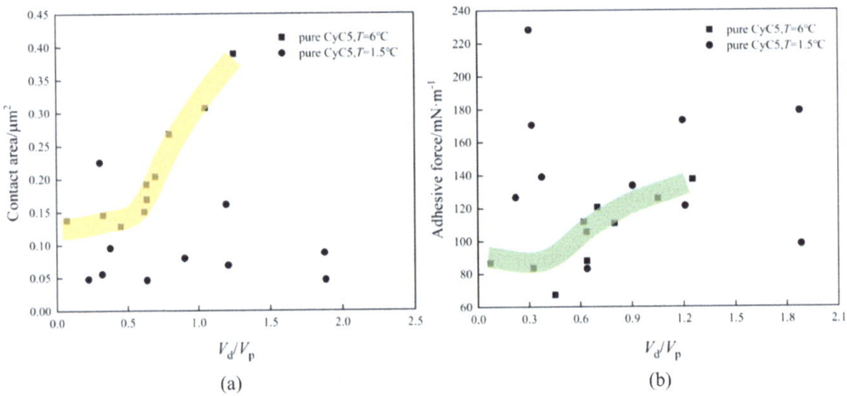

Fig. 5.15 Effects of the water volume on the contact area and the normalized adhesion force of hydrate particle-droplet in pure CyC5 at 1.5 °C and 6 °C, **a** Contact area, **b** Adhesive force

the hydrate particle, the water molecules around the TPC line would rapidly form additional hydrates under a higher subcooling/driving force, preventing the effective wetting of water on the hydrate particle surface and resulting in a smaller contact area. Also, the effects of the water volume on the contact area are disturbed by the high rate of hydrate formation. For the adhesion forces, they are also scattered. Furthermore, although the contact areas at 1.5 °C are significantly smaller than those at 6 °C, the corresponding adhesion forces at 1.5 °C are larger than those at 6 °C due to hydrate formation from the liquid water.

Compared to the adhesion force of hydrate particles, it has been found that the interaction force between hydrate particles and droplets is more than 10 times greater than those between hydrate particles, which is also verified by Aspenes [22]. In a word, the capillary adhesion force between particles would be significantly promoted by droplets and increase with hydrate formation. Besides, in the system without anti-agglomerant, the hydrate particle-droplet interaction plays a decisive role in the hydrate aggregation.

5.1.3 Interaction Force of Hydrate Particle-Droplet-Hydrate Particle Considering Liquid Bridge Solidification

5.1.3.1 Liquid Bridge Force Model

Currently, research on the model of hydrate particle-droplet interaction considering hydrate growth has never been described. Based on this, we attempt to establish a hydrate particle-droplet interaction model ("liquid bridge force model") considering hydrate growth/liquid bridge solidification. The typical schematic of hydrate particle-droplet interaction at a certain time is shown in Fig. 5.16. Here, the new TPC line induces the pendular liquid bridge between a hydrate sphere particle and a flat plate to shift to the pendular liquid bridge between a solid hydrate circular ring/platform and a flat plate. Therefore, the corresponding capillary force can be written as

$$F = \pi y(0)_t^2 \Delta P - 2\pi y(0)_t \gamma_{\text{hw}} \sin\theta_1 \tag{5.3}$$

where θ_1 shows the contact angle of liquid bridge with hydrate particle surface; $y(0)_t$ stands for the intersect point of the water-hydrocarbon-hydrate particle; ΔP represents the pressure difference (i.e., Laplace pressure) across the hydrocarbon/water

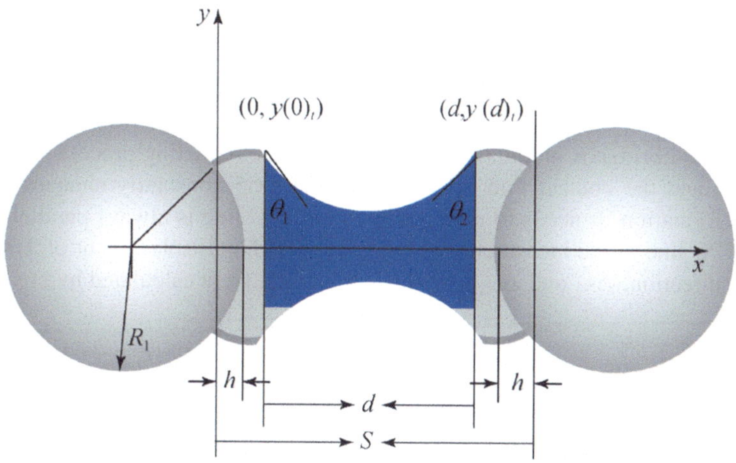

Fig. 5.16 Typical schematic of hydrate particle-droplet interaction at a certain time considering liquid bridge solidification

interface and could be calculated by Eq. 5.4. Where y' and y'' are the first and second derivatives of the liquid bridge profile described by the function $y(x)$ respectively; γ_{hw} refers to the interfacial tension of water and hydrocarbon. With the coordinates of the boundary points, ΔP can be calculated with Eq. 5.5.

$$\frac{\Delta P}{\gamma_{hw}} = \frac{1}{y\left(1 + y'^2\right)^{\frac{1}{2}}} - \frac{y''}{\left(1 + y'^2\right)^{\frac{3}{2}}} \tag{5.4}$$

$$\frac{\Delta P}{2\gamma_{hw}} = \frac{y(0)_t \sin(\theta_1) - y(d)_t \sin(\theta_2)}{y(0)_t^2 - y(d)_t^2} \tag{5.5}$$

5.1.3.2 Liquid Bridge Profile and Its Evolution

Liquid bridge profile is one key factor in determining the Laplace pressure of ΔP, and it could be described by a second-order polynomial shown in Eq. 5.6, where a, b and c are unknown constants. The apparent solid-liquid contact angles of θ_1 and θ_2 can be related to the liquid bridge profile shown in Eq. 5.7. For the initial contact, the contact angles could be written as Eqs. 5.8 and 5.9.

$$y(x) = ax^2 + bx + c \tag{5.6}$$

$$\begin{cases} \theta_1 = \frac{\pi}{2} + \tan^{-1}\left[y'(0)_t\right] \\ \theta_2 = \frac{\pi}{2} - \tan^{-1}\left[y'(d)_t\right] \end{cases} \tag{5.7}$$

$$\theta_1 = \frac{\pi}{2} + \tan^{-1}\left[y'(0)_t\right] - \sin^{-1}\left[\frac{y(0)_t}{R_1}\right] \tag{5.8}$$

$$\theta_2 = \frac{\pi}{2} - \tan^{-1}\left[y'(0)_t\right] - \sin^{-1}\left[\frac{y(d)_t}{R_2}\right] \tag{5.9}$$

The liquid bridge is gradually stretched during the retracting process of hydrate particles, and the contact line of hydrate-water-CyC5 constantly changes due to hydrate formation. It needs to note that the new TPC line is always parallel to the initial TPC line in the measurement. The corresponding displacement of TPC line in the x-direction, Δx_1 and Δx_2, could be approximatively calculated by:

$$\int_0^{\Delta x_1} \sqrt{1 + y(x)_t'^2}\,\mathrm{d}x = v_h(t) \cdot \mathrm{d}t \tag{5.10}$$

$$\int_{d_t - \Delta x_2}^{d_t} \sqrt{1 + y(x)_t'^2}\,\mathrm{d}x = v_h(t) \cdot \mathrm{d}t \tag{5.11}$$

where $\nu_h(t)$ represents the hydrate formation rate along the liquid bridge surface and could be determined experimentally from the bridge image analysis. It is worth noting that $\nu_h(t)$ is a strong function of the hydrate guest species and subcooling degree.

Suppose that $\nu_d(t)$ is the departure rate of hydrate particles, the liquid bridge length could be expressed as Eq. 5.12. Then the shape of the liquid bridge at $t + dt$ could be solved approximatively by Eq. 5.13. At $t = 0$, it could be expressed as Eqs. 5.14, 5.15 and 5.16.

$$d_{t+\mathrm{d}t} = d_t + \nu_d(t)\cdot \mathrm{d}t - \Delta x_1 - \Delta x_2 \tag{5.12}$$

$$\begin{cases} y(x)_{t+\mathrm{d}t}\big|_{x=0} = a(t)\cdot\Delta x^2 + b(t)\cdot\Delta x + c(t) \\ \pi\int_0^{d_{t+\mathrm{d}t}} y(x)^2_{t+\mathrm{d}t}\cdot \mathrm{d}x = \pi\int_{\Delta x_1}^{d_t-\Delta x_2} y(x)^2_t\cdot \mathrm{d}x \\ y(x)_{t+\mathrm{d}t}\big|_{x=d_{t+\mathrm{d}t}} = a(t)\cdot(d_t-\Delta x_2)^2 + b(t)\cdot(d_t-\Delta x_2) + c(t) \end{cases} \tag{5.13}$$

$$\pi\int_0^{d_0} y(x)_0^2\cdot \mathrm{d}x = V_{\mathrm{lip}}(0) + V_{\mathrm{cap1}}(0) + V_{\mathrm{cap2}}(0) \tag{5.14}$$

$$V_{\mathrm{cap1}}(0) = \frac{\pi h_0}{6}\left[3y(0)_0^2 + h_0^2\right] \tag{5.15}$$

$$V_{\mathrm{cap2}}(0) = \frac{\pi h_\mathrm{d}}{6}\left[3y(d)_0^2 + h_\mathrm{d}^2\right] \tag{5.16}$$

where the liquid bridge and flat plate boundary conditions are fixed or can be determined experimentally from the bridge image analysis. $V_{\mathrm{lip}}(0)$ is the initial volume of the liquid bridge, which can be estimated from image analysis before the droplet contacts the hydrate particle; $V_{\mathrm{cap1}}(0)$, $V_{\mathrm{cap2}}(0)$ are the initial volume of the spherical cap on top and bottom hydrate particle that is submerged into the liquid bridge.

5.1.3.3 Liquid Bridge Rupture

The liquid bridge presents one bottleneck in the middle section and usually ruptures at its thinnest point of water stretching. Gravity has negligible effects on the small liquid volumes. Herein, the droplets on hydrate particles surface can be approximated in the form of a perfect sphere cap under interfacial tension. Figure 5.17 shows the spherical cap of the droplet on the hydrate particle after rupture. R_{L1} is the equivalent radius of a liquid droplet on particle; T_{L1} is the cap maximum height of droplets on hydrate particles. All these parameters of droplet cap can be solved by Equations as below:

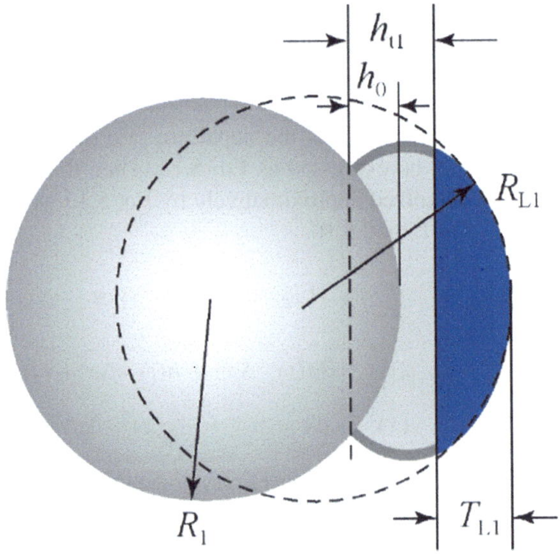

Fig. 5.17 Spherical cap of droplets on hydrate particle surface after liquid bridge rupture

$$\begin{cases} R_{L1} = \frac{T_{L1}^2 + y(0)_t^2}{2T_{L1}} \\ T_{L1}\left(3y(0)_t^2 + T_{L1}^2\right) = \frac{6\left[V_{L1}(t) + V_{cap1}(t)\right]}{\pi} \end{cases} \tag{5.17}$$

$$\begin{cases} V_{cap1}(t) = 0, h_{t1} \geq h_0 \\ V_{cap1}(t) = \frac{\pi}{6}(h_0 - h_{t1})\left[6R_1(h_0 - h_{t1}) - 2(h_0 - h_{t1})^2\right], h_{t1} < h_0 \end{cases} \tag{5.18}$$

$$\begin{cases} V_{cap2}(t) = 0, h_{t2} \geq h_0 \\ V_{cap2}(t) = \frac{\pi}{6}(h_d - h_{t2})\left[6R_2(h_d - h_{t2}) - 2(h_d - h_{t2})^2\right], h_{t2} < h_0 \end{cases} \tag{5.19}$$

$$\begin{cases} R_{L2} = \frac{T_{L2}^2 + y(d)_t^2}{2T_{L2}} \\ T_{L2}\left(3y(d)_t^2 + T_{L2}^2\right) = \frac{6\left[V_{L2}(t) + V_{cap2}(t)\right]}{\pi} \end{cases} \tag{5.20}$$

where h_{t1}, h_{t2} refer to the x-axial length of the top and bottom hydrate coating at the time of t, h_0 denotes the initial liquid bridge immersion depth; $V_{cap1}(t)$, $V_{cap2}(t)$ stand for the volume of the spherical cap on top and bottom of hydrate particle that is submerged into the pure liquid bridge at the time of t; T_{L1}, T_{L2} represent the respective water amount of liquid bridge distributed to the top and bottom of the hydrate particle.

Pepin criterion is proposed to predict the rupture of liquid bridge between two spheres cap, which assumes that the rupture occurs when the area of the liquid bridge before rupture is equal to the area of two droplets distributed on hydrate particles after rupture. This criterion will be proposed as following:

$$2\pi \int_{x=0}^{x=d_t} y(x)_t \sqrt{1 + y(x)_t'^2} \mathrm{d}x = 2\pi(R_{L1}T_{L1} + R_{L2}T_{L2}) \quad (5.21)$$

where the term on the left side is the interfacial area of the pure liquid bridge, while the term on the right side is the interfacial area of droplets.

Note that the liquid bridge always ruptures theoretically at the narrowest position. Therefore, the water volume of liquid bridge distributed on particle and plate could be calculated as

$$\begin{cases} V_{L1}(t) = \pi \int_{x=0}^{x=x_{\min}} y(x)_t \mathrm{d}x - V_{cap}(t) \\ V_{L2}(t) = \pi \int_{x=x_{\min}}^{x=d_t} y(x)_t \mathrm{d}x \end{cases} \quad (5.22)$$

Through the above model, the interaction force of hydrate particle-droplet-hydrate particle can be calculated. The comparison of the interaction force of hydrate particle-droplet-hydrate particle and hydrate particle-droplet-plate at different departure rates is shown in Fig. 5.18. The results show that their interaction force is the same without

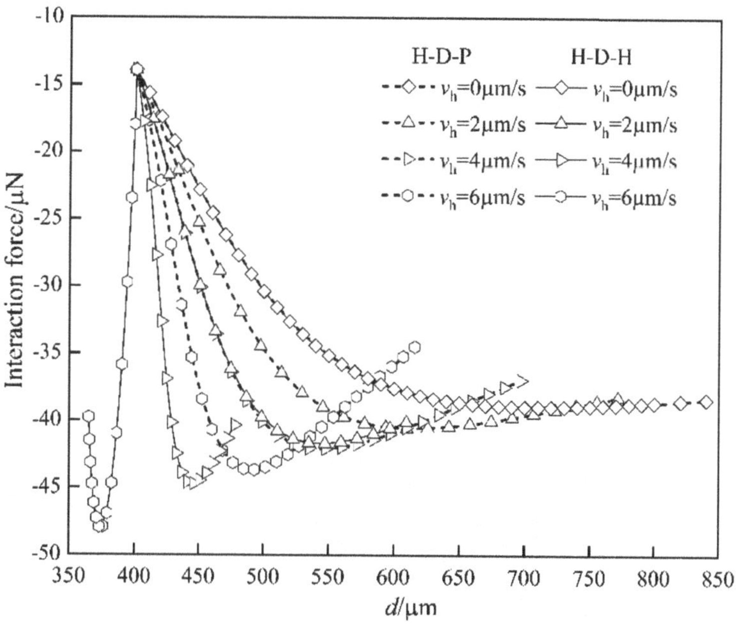

Fig. 5.18 Interaction force of hydrate particle-droplet-plate and hydrate particle-droplet-hydrate particle at different departure rates

considering hydrate formation, but the interaction force of hydrate particle-droplet-hydrate particles is evidently higher than that of hydrate particle-droplet-plate at the same rate of hydrate formation, which is also verified by the above measurement experiment. It needs to note that the length of the liquid bridge may be shortened with continuous stretching at a higher rate of hydrate formation.

5.2 Hydrate Deposition and Blockage Model in Gas-Liquid-Solid Three-Phase Flow

In gas-liquid-solid three-phase flow, liquid water is mainly in the form of liquid film on pipewall and droplets in the gas core. For the hydrate formation and deposition in this flow, the strong adhesion of hydrate layer converted from the liquid film could promote to attach the layer to the inner surface of the pipeline, and hydrate particles converted from droplets in the gas core could also be attached to the pipewall or migrated with flow [23], which is shown in Fig. 5.19.

Lorenzo et al. [24] carried out the hydrate formation and transportation experiment with E of 15.7%–18.8%. It is found that 30%–50% of hydrates in flow loop would be deposited on pipewall. In other words, 50%–70% of hydrates would migrate with gas and move away from the pipe. It needs to note that the hydrates departing from the pipe are mainly converted from the droplets in the gas core.

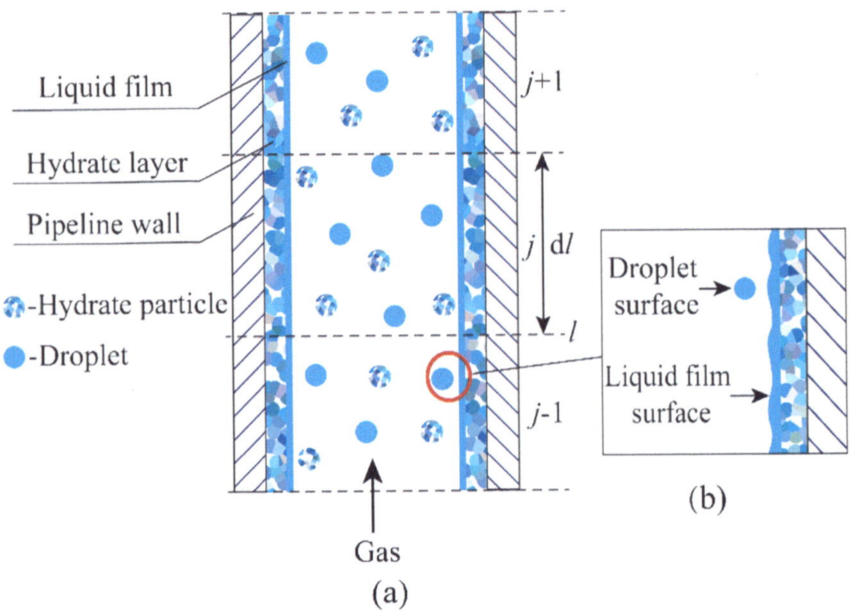

Fig. 5.19 Hydrate formation and deposition in gas-liquid-solid three-phase flow

Based on the hydrate formation and deposition experiment in flow loop.Wang et al. [25] established a hydrate deposition model considering the hydrate converted from the liquid film on pipewall. Through CFD simulation and related experiments, Jassim et al. [26, 27] found that for the individual hydrate particle, its deposition in the gas phase is significantly affected by factors such as the diameter of hydrate particle, the flow rate of gas, pipe inner diameter, wettability of pipewall, and a hydrate deposition model then is proposed for individual particle. However, in annular-mist flow, the hydrate deposition model considering the mass transfer and interaction of liquid film-droplets has never been described.

In this section, by introducing the droplets deposition theory in annular-mist flow and considering the mass transfer and interaction of liquid film-droplets, Wang et al. and Sun et al. [28] proposed a macro calculation method for hydrate particle deposition in the gas phase. The influence of hydrate deposition on the evolution of the hydrate layer on pipewall is analyzed thereby, laying the foundation for accurately predicting the hydrate blockage risks to some extent.

5.2.1 *Initial Deposition Model of Hydrate Particle in Gas Core*

For annular-mist flow, droplets in gas core and liquid film on pipewall are partly consumed and converted into hydrate, but their converting rates are different due to their different consumed amount of free water, breaking the equilibrium state of water content between droplets and liquid film. Afterwards, droplets deposition and liquid film atomization render the water content between them reach a new equilibrium. More specifically, droplets with larger Stokes numbers would migrate to the pipewall under inertia force. Schadel et al. [29] proposed that the deposition rate of droplets in gas core could be solved by measuring its flow as following:

$$R_{\mathrm{d}} = k_{\mathrm{d}}C = k_{\mathrm{d}}\left(\frac{W_{\mathrm{le}}}{Q_{\mathrm{g}}S}\right) = k_{\mathrm{d}}\left(\frac{4W_{\mathrm{le}}}{\pi D^2 U_{\mathrm{g}}S}\right) \tag{5.23}$$

where k_{d} represents the velocity of droplet migrating to pipewall, m/s; C is the concentration of droplet in gas core, kg/m^3; W_{le} shows the flow rate of droplets in gas phase, kg/s; Q_{g} shows the volume flow rate of gas, m^3/s; S refers to the slippage ratio, named the ratio between the velocity of droplets and gas; D is the pipe column diameter, m; U_{g} stands for the flow rate of gas, m/s.

In dl, the uniform distribution of droplets is assumed. Herein, the total amount of droplets depositing on pipewall at time dt is written as:

$$m_{\mathrm{ld}} = R_{\mathrm{d}}A_{\mathrm{l}} \cdot \mathrm{d}t = \pi k_{\mathrm{d}}C_{\mathrm{l}}D \cdot \mathrm{d}l \cdot \mathrm{d}t \tag{5.24}$$

where A_l shows the pipe wall area of dl, m^2; C_l is the concentration of droplets in the gas phase of dl, kg/m^3.

In annular-mist flow, the types of hydrate converted from droplets mainly comprise hydrate shells on droplet surface and fully converted hydrate particles. By now, the individual differences during the process of hydrate particle deposition are neglected. Assume that hydrate particles in dl are only the types of hydrate formation converted from droplets, and they are distributed uniformly and deposited on the pipewall with droplets deposition. Therefore, based on the droplets content in the gas phase (m_{le}), the amount of droplets deposition (m_{ld}) and hydrate content in the gas phase (m_{he}), the amount of hydrate particles depositing on pipewall in dl at time dt is expressed by Eq. 5.25.

$$m_{hd} = \frac{m_{he}}{m_{le}} m_{ld} \tag{5.25}$$

5.2.2 *Influence of Liquid Film Atomization on Hydrate Particles Deposition*

Liquid film atomization on pipewall could be described and explained well by the instability mechanism of Kelvin–Helmholtz. In other words, one rolling intermittent flow in the liquid film could be found owing to the shear force exerted by gas and adhesion force exerted by the pipeline wall. And the larger shear force, the higher flow rate of gas and low viscosity of liquid greatly promote the formation of rolling waves. When the ripples at the top of the rolling wave are keeping away from the surface of the liquid film, liquid film atomization occurs. Herein, in the rolling waves, hydrate particles depositing in the liquid film will be re-carried into the gas phase due to liquid film atomization which is shown in Fig. 5.20 [28].

The formation rate of hydrate converted from droplets is several times than that from the liquid film, causing more free water is consumed by hydrate converted

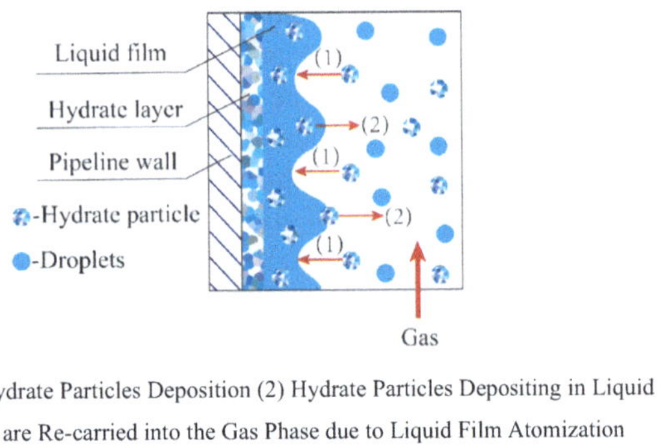

Fig. 5.20 Deposition process of hydrate particles in annular-mist flow

from droplets. The amount of liquid film atomization is far larger than the amount of droplet deposition during the equilibrium process of water content, so part of hydrate particles would be carried and returned to the gas phase by tiny droplets derived from liquid film atomization. Besides, the hydrates forming on the liquid film surface could prevent hydrate particles in the gas core from further contacting the liquid film. The adhesion force of hydrate particles-pipewall is reduced when hydrate particles immigrate to near the liquid film, so that the chances of hydrate particles depositing and attaching on pipewall are enormously remote [30]. In short, under the combined action of droplet deposition, liquid film atomization and hydrate formation on the liquid film surface, the possibility of hydrate particles depositing and attaching on pipewall is greatly reduced, causing the deposition amount of hydrate converted from droplet is much less than that from the liquid film.

Schadel et al. [29], Lopez de Bertodano et al. [31], Assad et al. [32] successively proposed and developed the calculation method for liquid film atomization in vertical pipes, which is expressed by Eq. 5.26.

$$R_{\mathrm{a}} = \frac{k_{\mathrm{a}}' U_{\mathrm{g}}^{2} (\rho_{\mathrm{g}} \rho_{\mathrm{l}})^{\frac{1}{2}}}{\sigma} \frac{W_{\mathrm{lf}} - W_{\mathrm{lfc}}}{P} \tag{5.26}$$

where k_{a}' stands for the nondimensional coefficient; P is the pipe perimeter, m; W_{lf} represents the mass flow of liquid film on pipewall, $W_{\mathrm{lf}} = W_{\mathrm{l}} - W_{\mathrm{le}}$, kg/s; W_{l} represents the mass flow of liquid, kg/s; W_{le} represents the mass flow of droplets, kg/s; W_{lfc} represents the critical mass flow of liquid film atomization, kg/s.

Thereby, the amount of droplet derived from liquid film atomization in dl at time dt is calculated as follows:

$$m_{\mathrm{la}} = R_{\mathrm{a}} A_{\mathrm{l}} \cdot \mathrm{d}t = \pi R_{\mathrm{a}} D \cdot \mathrm{d}l \cdot \mathrm{d}t \tag{5.27}$$

It can be seen that the amount of liquid film atomization is greatly influenced by the gas flow rate. The larger gas flow rate, the stronger rolling wave on the liquid film surface is, and the more droplets the liquid film atomization produces. Consequently, those droplets would immensely stop the hydrate particles in the gas core from depositing on pipewall and promote to carry the hydrate particles in the liquid film into the gas core.

5.2.3 *Effective Deposition Coefficient of Hydrate Particle in Gas Core*

Droplet deposition, liquid film atomization and hydrate formation on the liquid film surface interact with each other. In annular-mist flow, the free water of droplets in gas core and liquid film on pipewall are always in the dynamic equilibrium state of redistribution, making the deposition amount calculation of hydrate particles extremely difficult.

Only parts of hydrate particles are directly attached and bonded to the pipewall. However, either the rest are migrated and deposited to the liquid film on pipewall with droplet deposition or they eventually are carried out of pipe by gas with high flow rate. Herein, to investigate and represent the effect of liquid film atomization on the deposition of hydrate particles converted from droplets in the gas core, the parameter of "Effective Deposition Coefficient of Hydrate Particle in Gas Core (S_d)" is introduced [1, 28], which indicates in dl, the ratio between the amount of hydrate particles directly attached to the pipewall and the total amount of hydrate particles moving to the liquid film at a unit time. It is evident from the coefficient that S_d represents the actual deposition rate of hydrate particles in the gas-liquid-solid three-phase flow. The larger gas flow rate brings the greater E in the gas phase. Herein, the consumption rate of free water becomes faster due to the higher formation rate of hydrate particles, while the effect of liquid film atomization tends to be more obvious. Therefore, the droplets produced by liquid film atomization would immensely and strongly stop the hydrate particles in the gas core from depositing on pipewall. In addition, the higher formation rate of the hydrate layer converted from liquid film induces the smaller adhesion force between hydrate particles and pipewall, so it is difficult for hydrate particles to deposit and attach on pipewall. Consequently, this is creating the lessening of the. As aforementioned, S_d could be characterized as the function of subcooling degree of T_{sub}, Reynolds number of Re, the liquid flow volume of W_l, the surface tension of σ and Stokes value of S_t, which is shown as below:

$$S_d = f(T_{sub}, Re, W_l, \sigma, S_t) \tag{5.28}$$

The value of S_d under various experimental conditions (Table 5.1) could be obtained empirically according to 7 sets of experiments conducted by Lorenzo et al. [24]. It can be seen that the value of S_d is small and mainly distributed in range of 4%–6% caused by strong liquid film atomization. Therefore, the hydrate particle deposition model (Eq. 5.25) could be modified as follows:

Table 5.1 Value of S_d under various experiments conducted by Lorenzo

Experiment	1	2	3	4	5	6	7
Ethanediol concentration/%	0	0	0	10	20	20	20
Mean subcooling degree/°C	8.8	6.8	3.8	5.9	3.4	2.2	1.9
Reynolds number/10^4	121.6	119.9	118	119.6	123.3	123.9	122.2
Flow volume of liquid/L·min^{-1}	1.6	1.6	1.6	1.6	1.6	1.6	1.6
Surface tension/mN·m^{-1}	72.2	72.2	72.2	69.2	66.5	66.5	66.5
S_d	0.040	0.043	0.052	0.046	0.048	0.053	0.060

$$m_{\mathrm{hd}} = \frac{m_{\mathrm{he}}}{m_{\mathrm{le}}} m_{\mathrm{ld}} S_{\mathrm{d}} \tag{5.29}$$

In annular-mist flow, hydrate particles are constantly formed in the gas phase and deposited on the pipewall. The content of hydrate particles in the gas phase at present moment very much depends on the dimensions of the formation and deposition of hydrate particles at the previous moment, which is the sum of the hydrate particles that is not deposited and newly formed hydrate particles, indicating that hydrate content in gas phase is the function of time and space. By solving simultaneous Eqs. 5.23–5.29, the deposition amount of hydrate particles and newly formed hydrate particles in dl at time dt can be obtained, thereby the content of hydrate particles in dl could be written as follows:

$$m_{\mathrm{he},j+1}^{i+1} = m_{\mathrm{he},j}^{i} - \mathrm{d}m_{\mathrm{hd},j}^{i} + \mathrm{d}m_{\mathrm{he},j+1}^{i+1} \tag{5.30}$$

5.2.4 Hydrate Layer Growth and Hydrate Blockage

As the hydrate formation and deposition, the hydrate layer on pipewall would gradually become thickened, causing the effective cross-sectional area of the pipe to decrease, which is shown in Fig. 5.21. It indicates that the growth process of the hydrate layer on pipewall could be simply divided into three steps: (1) initial liquid film on pipewall, (2) hydrate film is converted from the liquid film, (3) hydrate layer appears as hydrate film is to become thicker continuously. Notably, hydrates that is located on pipewall is mainly derived from the hydrate layer converted from liquid film and deposited hydrate particles converted from droplets.

By using micromechanical force measurement, Nicholas et al. [33] and Aspenes et al. [22] found that the adhesion force of hydrate particles and pipewall are more than ten times as common due to the free water, which indicates that the hydrate layer converted from the liquid film would be probably deposited and attached to the pipewall, rather than being carried out of the pipe by gas or liquid flow. Herein, it

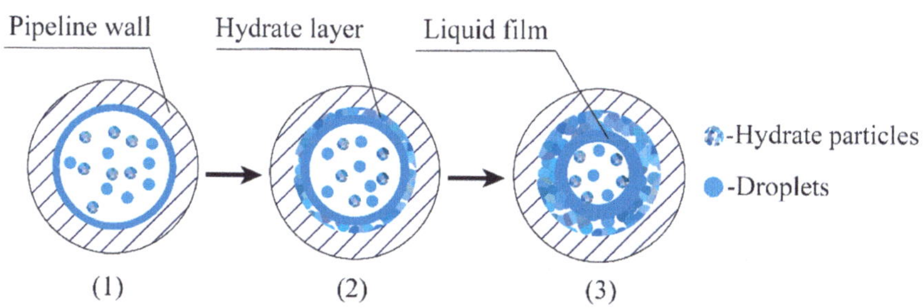

Fig. 5.21 Diagram of the growth of hydrate layer on pipewall

needs to suppose that all the hydrate layer converted from the liquid film is deposited and left inside of the pipe. Thus, in dl, the amount of hydrate layer converted from liquid film at time dt could be expressed as follows:

$$m_{\mathrm{hf}}=M R_{\mathrm{gf}}\mathrm{d}t \tag{5.31}$$

where m_{hf} shows the amount of hydrate layer converted from the liquid film, g; M is the molar mass of hydrate, g/mol; taking 119.5 g/mol for type II of hydrate; R_{gf} refers to the gas consumption rate during hydrate layer formation, mol/s.

Assume that the hydrate layer on pipewall is distributed uniformly and not detached from the pipewall in dl. Considering the hydrate converted from liquid film and hydrate particles converted from droplets in gas core constitute the whole amount of hydrate that is located on pipewall. So, the impact of hydrate deposition on δ is obtained and shown in Eq. 5.32 based on the Eqs. 5.29 and 5.31. For the dl, the change of pipeline inner diameter at time dt is written in Eq. 5.33.

Nevertheless, note that in most realistic cases, the growth rate of the hydrate layer is various at different positions of pipewall, leading to the non-homogeneous distribution of the hydrate layer. Moreover, the smaller pipeline inner diameter would enormously contribute and promote to grow the hydrate layer on pipewall, leading to the final hydrate blockage when the pipeline is fully occupied by hydrates.

$$\delta_{\mathrm{D}}=\frac{m_{\mathrm{hf}}}{\rho_{\mathrm{h}}\pi D\cdot \mathrm{d}l}+\frac{m_{\mathrm{hd}}}{\rho_{\mathrm{h}}\pi D\cdot \mathrm{d}l}=\frac{m_{\mathrm{hf}}+m_{\mathrm{hd}}}{\rho_{\mathrm{h}}\pi D\cdot \mathrm{d}l} \tag{5.32}$$

$$D_j^{i+1}=D_j^i-2\mathrm{d}\delta_{Dj}^i \tag{5.33}$$

5.2.5 *Model Solution and Validation*

Given the interaction of parameters among hydrate formation rate, hydrate particle concentration, the flow volume of liquid and inner diameter of pipeline, and their strong nonlinearity, the established model could be mathematically solved by finite difference method and their solution procedure is specifically expressed in Fig. 5.22 [1].

(1) Set interval dt, and divide the pipe column/pipeline into multiple-unit control bodies with a length of dl. Assume that the hydrate formation rate, δ_{D}, the concentration of droplet in the gas phase, hydrate particle concentration and hydrate deposition rate are generally distributed uniformly in dl;
(2) Calculate parameters such as the droplet entrainment fraction $E_{\mathrm{l},j}^i$, total surface area of the droplet $A_{\mathrm{d},j}^i$, inner surface area of pipe $A_{\mathrm{f},j}^i$ and velocity of droplet moving to the pipewall $V_{\mathrm{d},j}^i$ according to the initial parameters such as the flow

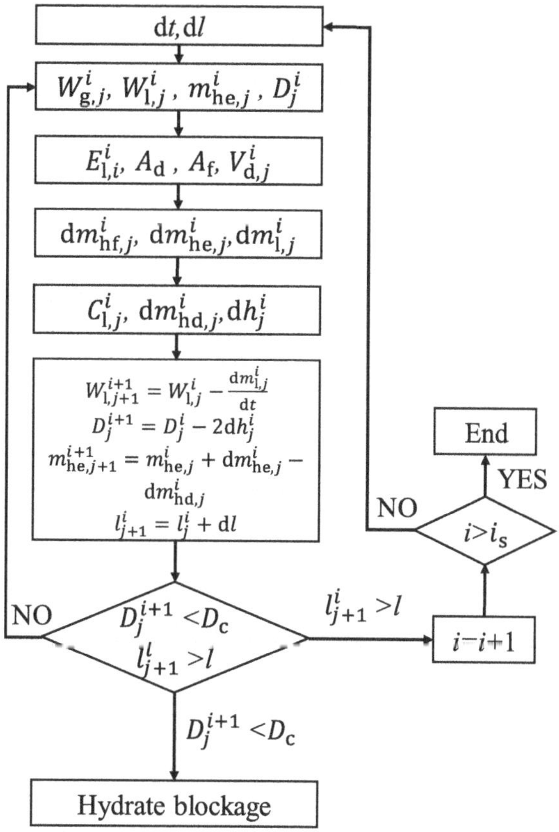

Fig. 5.22 Solution procedure for hydrate deposition and blockage model in gas-liquid-solid three-phase flow

volume of gas $W^i_{g,j}$, flow volume of liquid $W^i_{l,j}$, mass of hydrate particles $m^i_{he,j}$ and pipe inner diameter D^i_j in-unit control body j at the time i;

(3) Calculate the amount of hydrate layer converted from liquid film $dm^i_{hf,j}$, the number of hydrate particles converted from a droplet in gas core $dm^i_{he,j}$, the consumption amount of free water $dm^i_{l,j}$ respectively according to the temperature and pressure in the pipe, and obtain the variation regularity of parameters such as the concentration of droplet in the gas core $C^i_{l,j}$, the amount of hydrate particles deposition $dm^i_{hd,j}$, the hydrate layer thickness on pipewall $d\delta^i_{D_j}$ and pipe inner diameter D^{i+1}_j;

(4) Determine the initial parameters such as flow volume of liquid $W^{i+1}_{l,j+1}$, mass of hydrate particles $m^{i+1}_{he,j+1}$ and pipe inner diameter D^{i+1}_{j+1} in-unit control body $j+1$ at the time $i+1$ according to the previous calculation results in unit control body j at the time i;

(5) Calculate the initial parameters in unit control body $j+1$ at the time i according to the calculation results in unit control body j at time $i-1$ and obtain the time and

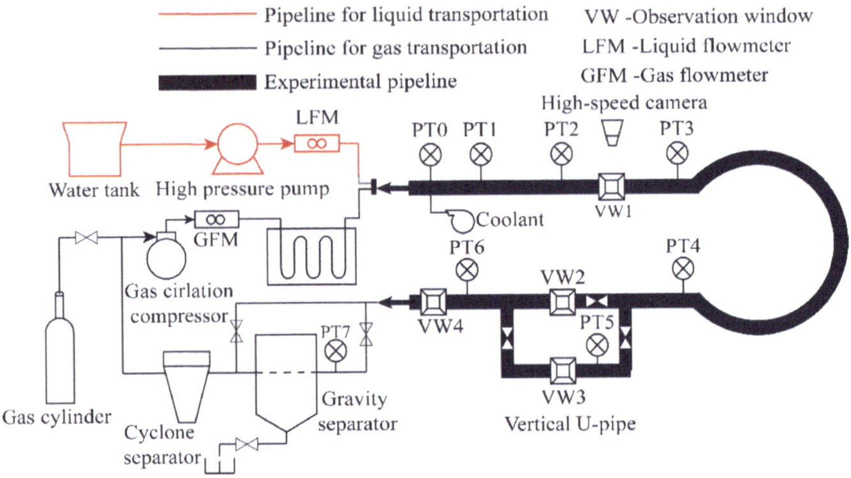

Fig. 5.23 Simplified layout of the flowloop conducted by Lorenzo [24]

space distribution of liquid flow volume, hydrate particle concentration, hydrate deposition amount, δ_D and pipe inner diameter through multiple iterations.

Eventually, hydrate blockage in the pipe would be found evidently when pipe inner diameter is smaller than the critical value D_c, which is the criteria to predict the specific occurrence time and position of hydrate blockage.

The laboratory experiment for hydrate formation and deposition in flow loop conducted by Lorenzo et al. [24] (Fig. 5.23) is selected and used to verify the hydrate deposition and blockage model. In this loop experiment, the flow loop (pipe inner diameter of 20.3 mm, pipe length of 40 m) is made up of stainless steel and one set of ethylene glycol cooling system with the minimum temperature being up to −8 °C is performed to regulate and control the pipewall temperature. The flow volume of gas (mixed natural gas) and liquid flow volume in the loop are 169 and 1.6 L/min with a respective apparent flow rate of 8.5 and 0.08 m/s. The value of S_d in gas phase is selected as 5%.

Based on the six groups of experimental conditions, the calculation results using the established model of hydrate deposition and blockage are shown in Table 5.2 and Fig. 5.24. It can be seen from the table that in dl, though the value of S_d at dt is only 5%, the number of hydrate particles deposited and attached to pipewall accounts for 24%–30% of the total hydrate particles in gas phase due to their continuous accumulation effect. From the figure, the relative errors of the hydrate formation rate in the loop pipe between theoretical calculation and the experimental result are within 10% and the average deviation is 6.73%. For the growth rate of the hydrate layer on pipewall, the calculation error is within 6% and the average deviation reaches 2.82%. The reasons will probably be because that the hydrate deposition and blockage model fail to fully consider the thermodynamic impacts of hydrate formation and deposition and the fact that the hydrate layer could be possibly detached and departed from the pipewall

Table 5.2 Results comparison of model calculation and loop experiment obtained by Lorenzo [24]

Experiment	Average subcooling degree/°C	Gas consumption/mol	Hydrate formation rate/L min^{-1}	Growth rate of hydrate layer/mm min^{-1}	Hydrate particles deposition ratio/%	Calculation deviation for hydrate formation/%	Calculation deviation for hydrate layer growth/%
1	3.1	100.85	0.318	0.064	30.42	9.72	1.89
2	3.4	111.17	0.338	0.071	30.46	8.71	3.15
3	4.2	127.06	0.386	0.067	26.23	1.03	5.88
4	4.5	117.09	0.357	0.069	24.25	8.05	0.62
5	4.6	119.54	0.364	0.071	24.35	4.08	0.36
6	5.5	110.10	0.365	0.068	24.09	8.82	5.03

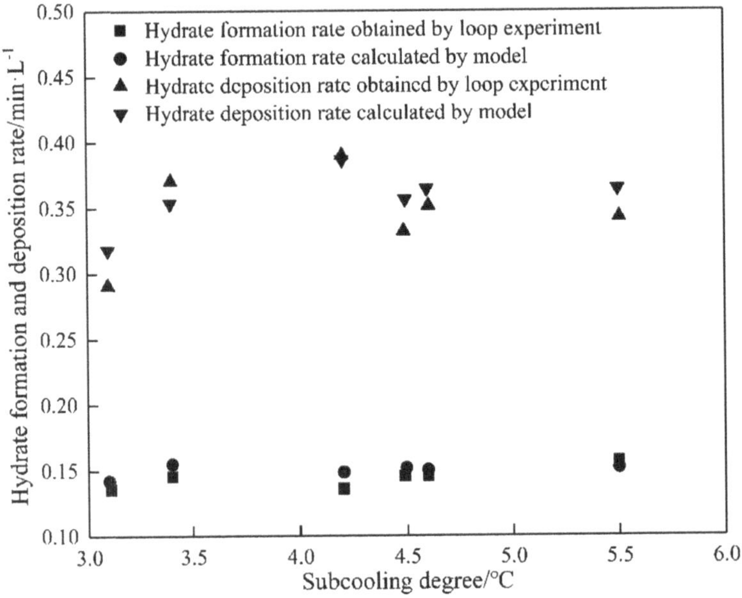

Fig. 5.24 Hydrate formation and deposition rate obtained by model and loop experiment conducted by Lorenzo [24]

is also neglected. It has also been found that the number of hydrates deposited and attached on the pipewall accounts for 38%–46% of the total hydrates in the pipe, indicating most of the hydrates would directly flow with gas, so that the status of three-phase flow is kept for a while.

Taken altogether, the results calculated by the established model agree quite well with the loop experiment, indicating that the hydrate deposition and blockage model could be available to help us predict the formation rate of hydrate particles and the growth rate of hydrate layer in annular-mist flow. Hence, when and where hydrate blockage occurs in the pipeline can be measured by rule and line.

The uniform pipe inner diameter [24] and change of pipe inner diameter at 41.5 min obtained by the model and is shown in Fig. 5.25, illustrating the maximum deviation of effective inner diameter is up to 20.4%, which could exert a great influence on other loop parameters. Under the condition of non-uniform effective inner diameter, the gas flow rate at different positions is shown in Fig. 5.26, illustrating the varied δ_D causes the gas flow rate to be different and such difference gradually increases with experimental time.

The uniform pipe inner diameter, as one important recognition criterion, is used to evaluate if the pipe is completely blocked by hydrates. Moreover, the criteria for judging the earliest time when hydrate blockage happens is the non-uniform effective inner diameter of the pipeline. Hence, the prediction time when hydrate blockage happens would be prolonged using the recognition criterion of uniform pipe inner

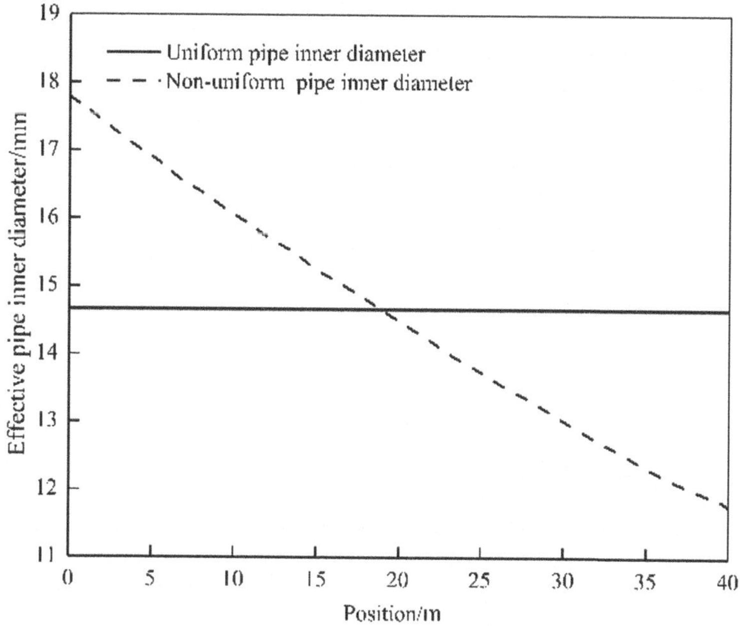

Fig. 5.25 Change of pipe inner diameter (41.5 min)

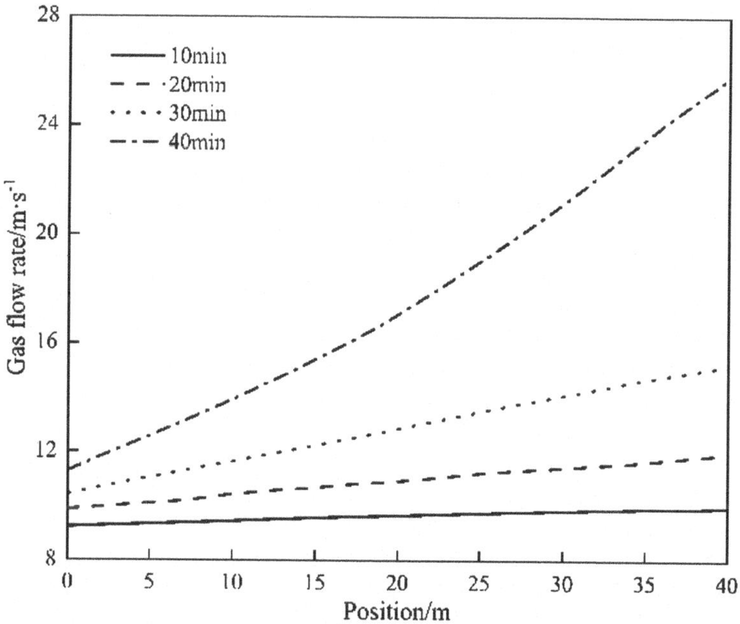

Fig. 5.26 Distribution of gas flow rate

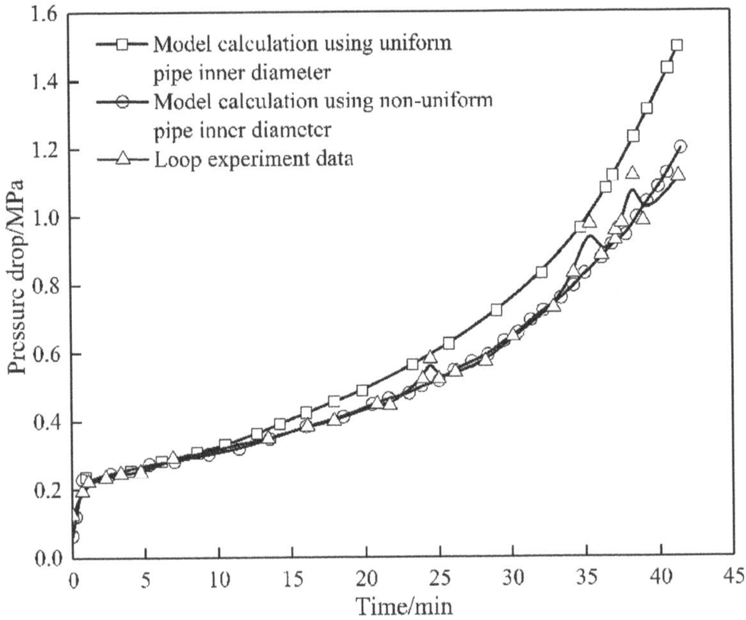

Fig. 5.27 Pressure drop obtained by model and loop experiment conducted by Lorenzo [24]

diameter, and the non-uniform effective inner diameter is more practical and suitable to predict this time.

The pressure drop change obtained by model and experimental data of Lorenzo [24] is shown in Fig. 5.27, showing that the pressure drop in an early stage of time is consistent with the experimental value using the uniform pipe inner diameter, but a larger deviation would appear with a long time. This is because in actual circumstances, the δ_D distribution is not uniform in pipeline, and its non-uniformity would be more evident with time.

The composition of the hydrate layer on pipewall is shown in Fig. 5.28, which illustrates that in the initial position, the deposition amount of hydrates converted from liquid film accounts for the vast majority of total deposition amount. Afterwards, the deposition amount of hydrates converted from droplets begins to increase along the pipeline and reaches 68% of total deposition amount at end of the pipeline due to the fact that at the front part of the pipeline, the formation amount of hydrates converted from droplets is greater than its deposition amount, making the hydrate particles concentration in the gas phase and its deposition amount gradually increase. This proves that the hydrate particle deposition in the gas phase is also an important factor causing the effective cross-sectional area of the pipeline to reduce, so this factor shall be taken into account in the hydrate blockage model. Otherwise, the actual δ_D in the pipeline would be larger than the prediction value, and corresponding plug removal would be easy to delay.

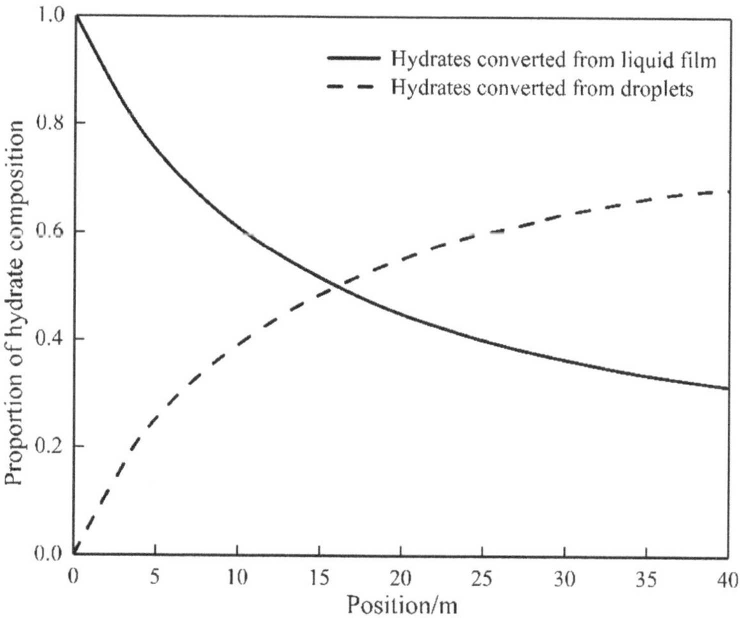

Fig. 5.28 Composition of hydrate layer on pipewall

From the above analysis, the S_d in the gas phase reflects the actual hydrate particles deposition to some extent. Based on the rate of hydrate formation and deposition at different positions of wellbore/pipeline, the hydrate blockage model considering S_d could be used to predict the time and position of hydrate blockage in advance.

5.3 Hydrate Deposition and Blockage Model in Gas-Solid Two-Phase Flow

The free water in the pipeline would gradually be consumed and converted into the hydrate particles or hydrate layer until no free water could be observed in the pipeline. Herein, the gas in the pipe would be pervasively under the water-saturated status, transforming the gas-liquid-solid three-phase flow into the gas-solid two-phase flow. However, the decreasing temperature of the pipeline is always developing due to the heat exchange between the gas and external low-temperature environment, and corresponding free water condensed from the gas then appears. Therefore, the growth of the hydrate layer on pipewall in gas-solid two-phase flow is all the result of the deposition of hydrate particles previously formed in the gas phase and hydrate particles converted from condensed water [34].

5.3.1 Theory on Radial Migration of Solid Particles in Gas-Solid Two-Phase Flow

Currently, the domestic and international studies on the hydrate particles migration in gas-solid two-phase flow are carried out mainly using the traditional migration kinetic theory of solid particles. Friedlander and Johnstone [35] conducted a lot of deposition experiments of the solid particle in turbulent flow and firstly proposed the Free-flight model to describe the migration of solid particles in the gas phase. Wells and Chamberlain [36] proposed a calculation method for the radial migration velocity of particles in gas-solid two-phase flow. Fichman et al. [37] established the radial migration model of solid particles considering the shearing action in the flow direction. Fan and Ahmadi [38] corrected the radial migration model considering the influence of two factors: the surface roughness and gravity of particles. Zhang and Zhou [39] established the particle orbit motion equation in gas-solid two-phase flow. Jassim et al. [26, 27] proposed a calculation method for the migration velocity of hydrate particles in radial and axial direction with the aid of the stress characteristics of the single hydrate particle in the pipeline. As shown in the aforementioned investigation, the radial migration velocity of solid particles in gas phase tends to present the "V-shape". Several calculation models for the radial migration velocity are briefly introduced as follows.

5.3.2 Wells and Friedlander Model

It is necessary as a starting point to think about "how is this model formed". The relaxation time of solid particles refers to the time scales of the migration velocity of particles adapted to the fluid velocity, which could be obtained by measuring the elapsed time when the initial velocity (0 m/s) of particles reaches the final motion velocity. Hence, the relaxation time of spherical particles could be expressed as

$$\tau_{\mathrm{v}} = \frac{u_{\mathrm{s}}}{g} = \frac{\rho_{\mathrm{p}} d_{\mathrm{p}}^{2} C_{\mathrm{c}}}{18\mu_{\mathrm{g}}} \tag{5.34}$$

Besides, the stopping distance of particles represents the length scale of particles keeping their original motion before they reach equilibrium status again in the variation flow field. And the stopping distance of particles is a linear function of the relaxation time, which could be expressed as Eq. 5.35, where V_0 is the initial velocity of solid particles.

$$S_{\mathrm{L}} = V_0 \tau_{\mathrm{v}} \tag{5.35}$$

For most of the solid particles, their motion in turbulent flow is mainly influenced by Brownian diffusion motion if $\tau_{\mathrm{v}}^{+} = \frac{\tau_{\mathrm{v}}(u^*)^2}{\nu_{\mathrm{g}}} < 0.3$ or $d_{\mathrm{p}} < 1\,\mu\mathrm{m}$ is met. Thus, the radial migration velocity of particles could be obtained by empirical formulas (Eq. 5.38) proposed by Wells and Chamberlain [36]. Where ν_g is the kinematic viscosity of the gas, m^2/s; *Sc* shows the Schmidt number; *Re* shows the Reynolds number of gas, which could be calculated by Eq. 5.37.

$$\frac{V_d}{U*} = 0.2Sc^{-2/3}Re^{-1/8} \tag{5.36}$$

$$Re = \frac{U_g D_{pipe}}{\nu_g} \tag{5.37}$$

On the other hand, the motion of particles in turbulent flow is mainly influenced by inertia force if $\tau_v^+ > 0.3$ or $d_p < 1\mu m$ is met. Herein, The radial migration velocity of particles could be obtained by empirical formulas (Eq. 5.38) proposed by Friedlander and Johnstone [35]. Where *F* shows the Moody's friction resistance coefficient; S_L^+ could be expressed by Eq. 5.39.

$$V_{dr} = \begin{cases} \frac{f/2}{1+\sqrt{f/2}\left(1525/\left(S_L^+\right)^2-50.6\right)}, S_L^+ < 5 \\ \frac{f/2}{1+\sqrt{f/2}\{5\ln[5.04/\left(S_L^+/5-0.965\right)]-13.73\}}, 5 \le S_L^+ \le 30 \\ \frac{f}{2}, 30 < S_L^+ \end{cases} \tag{5.38}$$

$$S_L = \frac{\rho_g \rho_p d_p U_g f}{36\mu_g} \tag{5.39}$$

5.3.2.1 Wood Model

Wood et al. [40] introduced the ratio of radial migration velocity to friction velocity to express the dimensionless radial migration velocity of solid particles in turbulent flow, which is shown in Eq. 5.40, while the corresponding empirical formula based on many experiments is shown in Eq. 5.41. Where the first item on the right represents the impact of Brownian motion; the second item on the right represents the eddy diffusion effect, and the third item represents the radial migration effect caused by particle gravity in the horizontal pipeline.

$$V_{dr}^+ = \frac{V_{dr}}{u*} \tag{5.40}$$

$$V_{dr}^+ = 0.057Sc^{-2/3} + 4.5 \times 10^{-4}\tau^+ + \tau^+ g^+ \tag{5.41}$$

$$\tau^+ = \frac{\tau u^{*2}}{\nu} = \frac{\rho_p d_p^2 C_c u^{*2}}{18\rho_f \nu^2} \tag{5.42}$$

$$g^{+} = \frac{\nu g}{u^{*2}} \tag{5.43}$$

5.3.2.2 Fan and Ahmadi Model

When the surface roughness of pipewall and particle gravity are taken into account, Fan and Ahmadi [38] proposed the empirically modified formulas (Eq. 5.44) to calculate the radial migration velocity of solid particles in the horizontal pipeline. Where u^{*} refers to the friction velocity, m/s; k^{+} shows the surface roughness and $k^{+} = 0$ when the pipewall has a smooth surface.

$$V_{\mathrm{dr}} = \begin{cases} 0.084Sc^{-2/3} + 0.5\left[\dfrac{\left(0.64k^{+} + 0.5d_{\mathrm{p}}^{+}\right)^{2}}{3.42}\right]^{1/\left(1+\tau^{+2}L_{1}^{+}\right)} \times \left[1 + 8\mathrm{e}^{-(\tau-10)^{2}/32}\right]\dfrac{0.037}{1-\tau^{+2}L_{1}^{+}} \\ 0.14, \text{ if } V_{\mathrm{dr}} \geq 0.14 \end{cases} \tag{5.44}$$

$$V_{\mathrm{dr}} = \frac{V_{\mathrm{dh}}}{u^{*}} \tag{5.45}$$

$$L_{1}^{+} = \frac{3.08}{Sd_{\mathrm{p}}^{+}} \tag{5.46}$$

$$S = \frac{\rho_{\mathrm{p}}}{\rho_{\mathrm{f}}} \tag{5.47}$$

$$d_{\mathrm{p}}^{+} = \frac{d_{\mathrm{p}}u^{*}}{\nu} \tag{5.48}$$

5.3.2.3 Jassim Model

Sosnowski et al. [41] found that the motion of tiny particles in the gas phase mainly depends on the aerodynamic parameters, while the motion of larger particles is mainly influenced by their motion inertia. To better describe the particle's motion characteristics in the area of Brownian motion, Jassim et al. [27] proposed one modified formula for the deposition velocity of particles (Eq. 5.49) on the basis of introducing the Wells model, Fan and Ahmadi model into the Wells model. Comparing the calculation results of the four models above, experimental data obtained by Kvasnak et al. [42] and simulation results obtained by Tian et al. [43] and Shams et al. [44], Jassim concluded that the Jassim model had a better effect on descriping the motion characteristics of solid particles in the gas phase.

$$V_{\mathrm{dr}} = \begin{cases} 0.2Sc^{-2/3}Re^{-1/8} + 0.5\left[\dfrac{\left(0.64k^{+}+0.5d_{\mathrm{p}}^{+}\right)^{2}}{3.42}\right]^{1/\left(1+\tau^{+2}L_{1}^{+}\right)} \times \left[1+8\mathrm{e}^{-(\tau-10)^{2}/32}\right]\dfrac{0.037}{1-\tau^{+2}L_{1}^{+}} \\ 0.14, \text{ if } V_{\mathrm{dr}} \geq 0.14 \end{cases} \tag{5.49}$$

5.3.3 *Hydrate Particles Deposition in Gas-Solid Two-Phase Flow*

In gas-solid two-phase flow, the hydrate particles in the gas phase are gradually deposited and attached on pipewall until no particles could be observed in the pipeline [1]. Currently, the theory on the solid particle deposition in gas-solid two-phase flow is frequently used to investigate the deposition of hydrate particles in the gas phase. Gu and Xu [45] established one particle concentration model in turbulent flow for calculating the motion velocity and concentration of particles in gas-solid two-phase flow. Lu and Zhao [46] simulated and studied the gas-solid two-phase flow in pipe based on the gas-solid two-phase flow model, turbulence flow model of particles and gas phase. Jassim et al. [27] found that the deposition rate of hydrate particles is chiefly affected by their diameter. If the diameter of hydrate particles is larger than the critical value, the deposition rate will be not going to change. A new hydrate particle deposition model is established accordingly, which divides the motion area of hydrate particles into turbulent flow area and boundary layer area. In the turbulent flow area, the motion velocity of hydrate particles moving to the pipewall could be calculated by Eq. 5.49, which determines the lasting time that particles reach the boundary layer and the size of the boundary layer as well. However, in the boundary layer area, the main influencing factors for the subsequent motion of hydrate particles are their size and motion velocity. Hydrate particles would be likely prone to enter the boundary layer until the particle size is larger than the boundary layer thickness, and the motion velocity of hydrate particles then mainly determine their motion state. Herein, the hydrate particles would have a bounce-back effect if their motion velocity is greater than the critical velocity. Instead, the attachment and deposition of the particles would occur. According to the energy conservation before and after hydrate particles bounce, Friedlander [35] proposed the Eq. 5.50 for calculating the motion velocity after particles bounce, and the critical velocity could be written as Eq. 5.51 when $v_2 = 0$. Where m_{p} shows the particle mass, kg; e represents the recovery coefficient, dimensionless.

$$\frac{v_2}{v_1} = \left[e^2 - \frac{E\left(1-e^2\right)}{m_{\mathrm{p}}e^2/2}\right]^{1/2} \tag{5.50}$$

$$v_{1\mathrm{c}} = \left[\frac{2E\left(1-e^2\right)}{m_{\mathrm{p}}e^2}\right]^{1/2} \tag{5.51}$$

To date, the hydrate particle deposition model mentioned above is mostly used to describe the deposition of individual hydrate particles in the gas phase, but the deposition model for the hydrate particle group in the gas phase is rarely reported. For the hydrate particle group, it could be found that the deposition rate of hydrate particles mainly depends on the particles concentration in gas phase and velocity of hydrate particles moving toward pipewall by taking advantage of the theory on radical migration of solid particles effectively in gas-solid two-phase flow [47].

Assume that the concentration distribution of hydrate particles in gas phase is even in dl. Herein, the hydrate particle deposition rate could be expressed by Eq. 5.52. Where V_{dh} shows the velocity of hydrate particles moving toward pipewall, m/s; C_{he} refers to the hydrate particles concentration at different positions of pipeline, which could be calculated by Eq. 5.53; V is the volume of dl, m^3; m_{he} is the amount of hydrate particles in dl at different positions of the pipeline, kg.

$$R_{\text{dh}} = V_{\text{dh}} C_{\text{he}} \tag{5.52}$$

$$C_{\text{he}} = \frac{m_{\text{he}}}{V} \tag{5.53}$$

The adhesion force between hydrate particles and pipeline wall is the key factor influencing the attachment and deposition of hydrate particles. By measuring the adhesion force between CyC5 hydrate particles and pipewall, Nicholas et al. [33] and Aspenes et al. [22] found that free water on pipewall had a great influence on the adhesion force. More particularly, when the pipe wall is wet, the adhesion force between hydrate particles and pipewall was greater than that between hydrate particles. However, one opposite result in dry pipewall could be obtained, indicating it is more difficult for hydrate particles in the gas phase to attach and deposit on pipewall without free water than with free water, and the effective deposition coefficient of hydrate particles without free water is getting smaller. In the gas-liquid-solid three-phase flow shown in Sect. 5.2, only about 5% of hydrate particles moving toward pipewall would be deposited on pipewall, and the rest would be carried back to the gas phase due to liquid film atomization. From this, it could be suggested that, in gas-solid two-phase flow, only a few hydrate particles would be deposited under the smaller adhesion force between particles and pipewall. The motion of hydrate particles having been not deposited mainly includes three types: sliding, rolling and bounce.

Herein, the effective deposition rate of hydrate particles in gas-solid two-phase flow is commonly represented by S_{d2}, which could be expressed by Eq. 5.54 [1]. According to mass conservation, the amount of hydrate particles at a different time and position is obtained by Eq. 5.55. The growth rate of the hydrate layer on pipewall could be calculated by Eq. 5.56 in terms of the corresponding hydrate particles deposition rate changing with time and position. Therefore, the δ_{D} at one position of pipewall and at a given time could be calculated by Eq. 5.57. Where G is the growth rate of the hydrate layer on pipewall, m/s; ρ_{h} shows the hydrate density, kg/m^3; δ_{D} stands for the hydrate layer thickness on pipewall, m; t is the time, s.

$$R_{\mathrm{dh}} = V_{\mathrm{dh}} C_{\mathrm{he}} S_{\mathrm{d2}} \tag{5.54}$$

$$m_{\mathrm{he},j+1}^{i+1} = m_{\mathrm{he},j}^{i} - \mathrm{d}m_{\mathrm{hd},j}^{i} \tag{5.55}$$

$$G = \frac{M R_{\mathrm{dh}}}{\rho_{\mathrm{h}} \pi D \mathrm{d}l} = \frac{M V_{\mathrm{dh}} C_{\mathrm{he}} S_{\mathrm{d2}}}{\rho_{\mathrm{h}} \pi D \mathrm{d}l} \tag{5.56}$$

$$\delta_{\mathrm{D}} = \int_0^t G \mathrm{d}t = \int_0^t \frac{M V_{\mathrm{dh}} C_{\mathrm{he}} S_{\mathrm{d2}}}{\rho_{\mathrm{h}} \pi D \mathrm{d}l} \mathrm{d}t \tag{5.57}$$

5.3.4 Model Solution and Validation

5.3.4.1 Model Solution

In gas-solid two-phase flow, the interrelationship between hydrate particles concentration, hydrate particles deposition rate, δ_{D} and the effective inner diameter of pipe undoubtedly leads to the strong nonlinearity of hydrate deposition and blockage model. Its numerical solution could be solved with the finite difference method and the solution procedures are as follows [1]:

(1) Set small enough interval dt and divide the pipeline into multiple-unit control bodies with a length of dl. In dl, suppose that: ① Both droplets in the gas phase and hydrate particle concentration is distributed uniformly, so that the hydrate deposition rate at different positions is the same; ② The heat transfer only occurs at the radial direction of pipe column; ③ The hydrate density remains constant and the porosity of hydrate layer on pipewall is neglected.

(2) Calculate the radial migration velocity and deposition rate of hydrate particles when the initial amount of hydrate particles (m_0) at the inlet of first dl is given according to Eqs. 5.49 and 5.53; calculate the amount of hydrate particles migrating to the second unit control body (m_2) according to mass conservation theory. Followed by analogy, obtain the amount and deposition rate of hydrate particles in unit control body i.

(3) Calculate the growth rate of the hydrate layer in different unit control bodies according to Eq. 5.55, and calculate the δ_D at different positions of pipeline at one time by Eq. 5.56.

(4) Repeat the above steps to calculate the hydrate particles amount, deposition rate of hydrate particles, and δ_{D} in next dl when the length l_{i} is smaller than the length of pipe column; otherwise, calculate δ_{D} at different positions of pipe next moment until the calculation time is required.

5.3.4.2 Model Validation

In this chapter, the hydrate deposition and blockage model is validated through the deposition experiment of individual hydrate particles in flow loop conducted by Jassim et al. [27]. Diagram for the hydrate deposition experiment is shown in Fig. 5.29. In this flow loop, pipe inner diameter is 19.05 mm, and the experimental temperature and pressure remain 5.5 °C and 0.193 MPa respectively.

For a single hydrate particle, the deposition distance obtained from experiments and model calculation results is shown in Table 5.3. it is can be found that the deviation between model calculation and experimental results is small, indicating the hydrate deposition and blockage model based on the theory on hydrate particle radial migration has high accuracy on prediction of hydrate deposition.

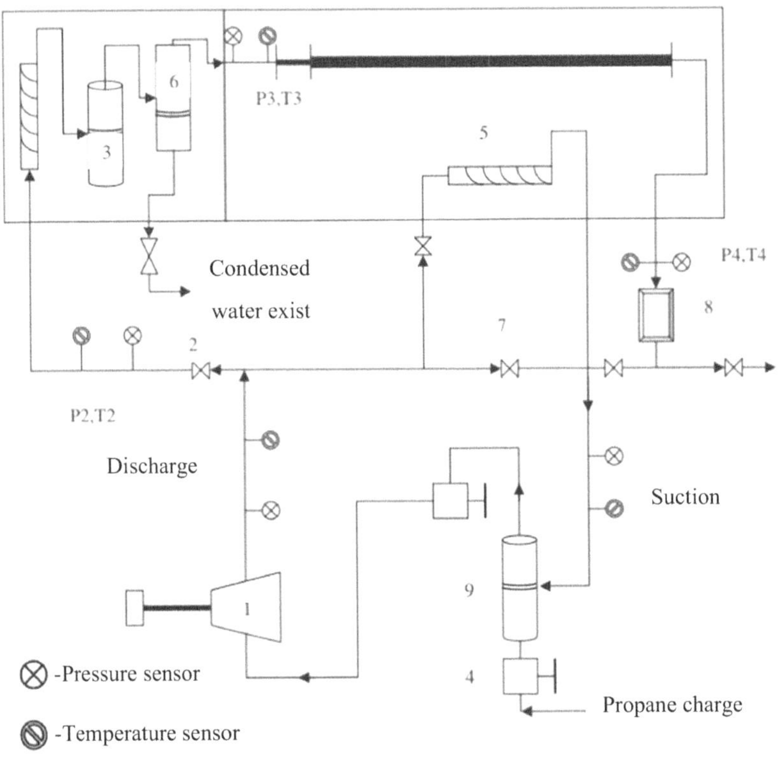

1—Compressor; 2—Valve; 3—Gas saturator; 4—Gas charging valve; 5—Cooling coil; 6—Separator; 7—Valve; 8—Flowmeter; 9—Liquid trap/filter

Fig. 5.29 Schematic diagram for hydrate deposition experiment in flowloop conducted by Jassim [27]

Table 5.3 Deposition distance of hydrate particle obtained by loop experiment and model calculation

S N	*Re*	Deposition distance measured by loop experiment/m	Deposition distance calculated by the established model/m	Relative deviation/%
1	4300	0.746	0.735	1.5
2	8400	0.823	0.837	1.7
3	14500	0.891	0.905	1.6

Further, for the model validation, suppose that the hydrate particle concentration at inlet reaches 5.13 kg/m^3; pipe inner diameter is 114.3 mm, and the flow rate of gas is 4.86 m/s. Under this condition, the sensitivity analysis on the effective deposition coefficient of hydrate particles (EDR_{g-s}) in gas-solid two-phase flow will be performed.

The relationships between EDR_{g-s} and the deposition distance of hydrate particles is shown in Fig. 5.30, showing the EDR_{g-s} has a significant influence on the deposition distance in gas-solid two-phase flow. Specifically, the larger EDR_{g-s} induces the smaller particles deposition distance. The quantitative change location of flow pattern shifting from the gas-solid two-phase flow to the water-saturated gas single-phase flow becomes apparently closer to the inlet of pipeline.

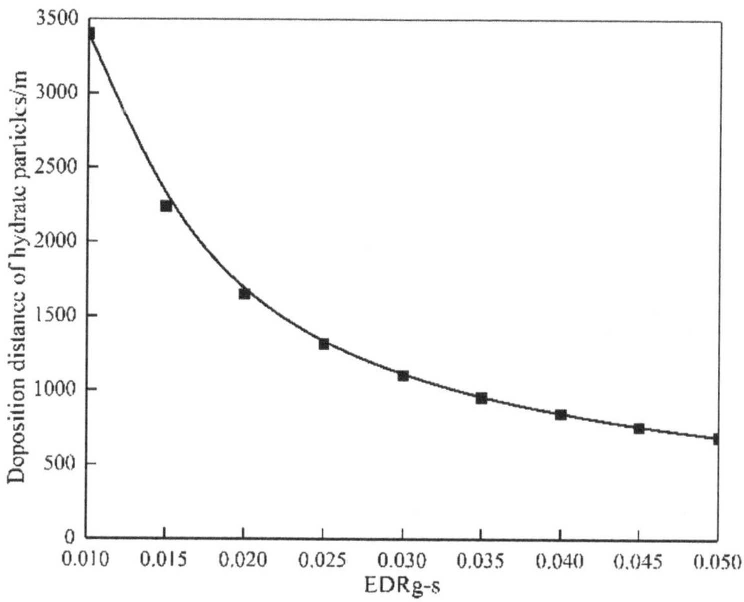

Fig. 5.30 Sensitivity analysis of EDR_{g-s}

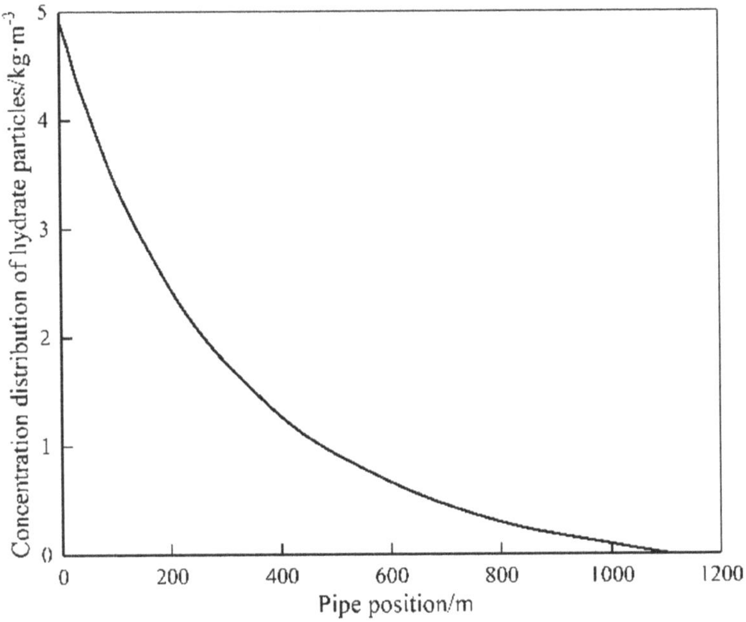

Fig. 5.31 Concentration distribution of hydrate particles

So far, the study on the $EDR_{g\text{-}s}$ in gas-solid two-phase flow has not been reported yet. Herein, the value of $EDR_{g\text{-}s}$ is set as 3% temporarily. Through the model calculation, the concentration distribution of hydrate particles in two-phase flow is shown in Fig. 5.31. It is can be found that, as the hydrate particles are continuously deposited, their concentration would gradually decrease along the pipeline until the overall hydrate particles are deposited on pipewall. The reason will mainly be because no new hydrate particles are formed and supplied to the gas phase during particle deposition.

5.4 Hydrate Deposition and Blockage Model in Water-Saturated Gas Single-Phase Flow

As hydrate particles are continuously deposited and attached on the pipewall in gas-solid two-phase flow, the amount of hydrate particles would gradually decrease along the pipeline. The water-saturated gas single-phase flow would then appear in subsequent flow if all the hydrate particles are completely deposited and attached at a certain position of the pipeline. However, some condensed water could still be separated from saturated gas flow due to the low temperature of pipewall, so that hydrate formation, deposition, and blockage will inevitably occur at high pressure and low temperature [1, 34]. Therefore, hydrate deposition in water-water-saturated gas systems is also an important aspect influencing the flow assurance in deepwater oil and gas development.

5.4.1 Hydrate Deposition and Blockage Model

In water-saturated gas systems, unlike the free water, the condensed water is always precipitated on the pipewall from water-saturated gas due to the low temperature of pipewall, and would be brought into contact with the gas to form the corresponding hydrate. Lingelem et al. [48] considered the hydrate growth was similar to the freezing process in the flow system with a continuous gas phase. The hydrate formation and deposition on pipewall could eventually form the hydrate film. Nicholas et al. [33] and Rao et al. [49] considered the hydrate deposition in saturated gas flow was similar to the process of wax deposition. Both these processes include: the dispersed phase spreading to solid wall, separation and then deposition. Note that the small amount of condensed water separated from water-saturated gas would lead to the low hydrate deposition rate. Hence, the required time will become longer when the hydrate layer converted from the condensed water is formed evidently in saturated gas flow. Rao et al. [49] found δ_D was only 2 mm when the process of particle deposition lasts for 60 h. Nicholas et al. [33] proposed the condensation process of saturated gas often occurred on the low-temperature surface where the gas was brought into contact with the pipewall, and believed that the hydrate converted from the condensed water was deposited and attached to the pipewall in place. Herein, it is salutary to consider that the hydrate deposition rate is equal to the hydrate formation rate, so the hydrate deposition rate could be illustrated by Eq. 5.58.

In saturated gas single-phase flow, hydrate formation and deposition equally obey the conservation law of mass and energy. Based on this, the effective inner diameter of pipe and temperature distribution at different positions of the pipeline could be obtained by Eqs. 5.59 and 5.60 respectively. The temperature and pressure at different positions of the pipeline will result in a different subcooling degree and the amount of condensed water, further leading to the various rates of hydrate formation and deposition. Therefore, the δ_D distribution at different positions of pipe turns to be non-uniform in water-saturated gas systems.

$$R_{dh} = R_{hf} = 2\pi r_i h_m [C_B - C_S] dl \tag{5.58}$$

$$2\pi \Delta z r_i h_m [C_B - C_i(T_i)] = \frac{d}{dt}\left[\pi \Delta z \left(r_{ti}^2 - r_i^2\right)\rho_h\right] \tag{5.59}$$

$$2\pi \Delta z r_i h_B (T_B - T_i) - 2\pi \Delta z r_i \mu\prime (T_i - T_c) + 2\pi \Delta z r_i h_m [C_B - C_i(T_i)] \Delta H_f = 0 \tag{5.60}$$

$$\mu \frac{1}{\ln({}^{r_{ti}}/_{r_i}) + K_{h+1}} \tag{5.61}$$

$$Nu = 0.023 Re^{4/5} Pr^{1/3} = \frac{h_B D}{k} \tag{5.62}$$

where h_m refers to the mass transfer coefficient, m/s; r_i shows the effective flowing inner radius of the pipeline after hydrate formation, m; r_{ti} shows the initial inner diameter of the pipeline, m; C_B and C_S are the water molecular concentration under gas temperature and pipewall temperature, kg/m^3; $C_i(T_i)$ is the water saturation in gas after hydrate formation, kg/m^3; Δz stands for the step length, m; ρ_h is the hydrate density, kg/m^3; ΔH_f represents the hydrate formation enthalpy, kJ/kg; $\mu\prime$ is the comprehensive heat transfer coefficient, which could be calculated by Eq. 5.61; h_B shows the internal convection coefficient of natural gas in the pipe, W/(m^2·K), calculated by Eq. 5.62; k_h shows the heat conductivity coefficient of hydrate, W/(m·K); h_c shows the heat transfer coefficient of seawater, W/(m^2·K); Nu is the Nusselt number; Re is the Reynolds number and Pr is the Prandtl number; D refers to the inner diameter of the pipeline, m; k represents the heat conductivity coefficient of natural gas, W/(m·K).

5.4.2 Model Solution and Validation

Given the interrelationship of parameters among temperature and pressure in the pipe, the saturation concentration of water molecules, hydrate deposition rate, δ_D, the effective inner diameter of the pipe and their strong nonlinearity in water-saturated gas systems, the established model could be mathematically solved with finite difference method and their solution procedure is specifically expressed as follow [1]:

(1) Set small enough interval dt and divide the pipeline into multiple-unit control bodies with a length of dl. In dl, suppose that: ① The temperature and pressure are distributed uniformly; ② The heat transfer only occurs at the radial direction of pipe column; ③ The hydrate density remains constant and the porosity of the hydrate layer on pipewall is neglected.
(2) Calculate the saturated vapor pressure and water saturations when the temperature and pressure at the inlet of first dl are known according to the above Equations.
(3) Calculate the temperature and pressure, the saturated vapor pressure and water saturations at the outlet of dl based on the temperature and pressure loss of saturated gas.
(4) Calculate the amount of condensed water and corresponding hydrate formation and deposition according to the water saturations difference at the inlet and outlet of dl, then obtain the δ_D according to mass conservation.
(5) Repeat the above steps to calculate the amount of hydrate particles, particle deposition rate and δ_D in next dl when the length l_i is smaller than the pipe column length; otherwise, calculate the δ_D at different positions of pipe next moment until the calculation time is required.

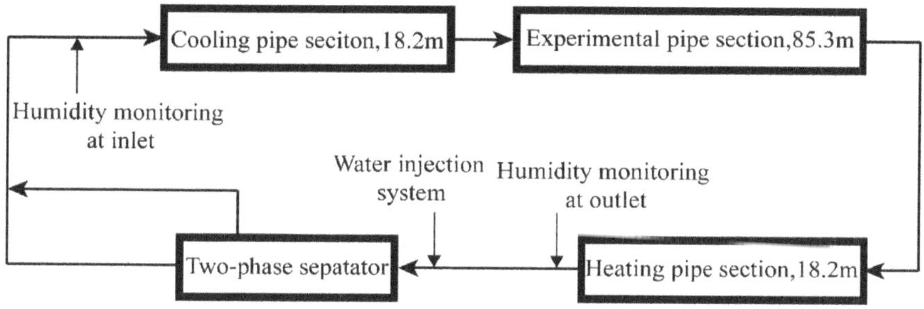

Fig. 5.32 Diagram of hydrate formation and deposition experiment in flowloop conducted by Nicholas [33]

For the saturated gas flow, Nicholas et al. [33] has conducted the hydrate formation and deposition experiment in one flow loop shown in Fig. 5.32, in which the inner diameter and length of the experimental pipe are 9.3 mm and 85.3 m respectively. By now, this loop experiment is used to validate the hydrate deposition and blockage model in saturated gas flow.

The changes of pressure drop and temperature obtained by the hydrate deposition and blockage model and the experimental measure are shown in Fig. 5.33. In this figure, the data calculated from the model has a good agreement with the experimental values and the percentage error between the model and experiment is not more than 10%. Moreover, the changes of water saturations at different temperatures and pressure are shown in Figure 5.34, which shows that the higher temperature and lower pressure in pipe will cause greater water saturation. It needs to notice that the water saturation changes caused by pressure are often ignored due to the fact that the saturated gas flow has greater temperature variation and smaller pressure variation in the pipe.

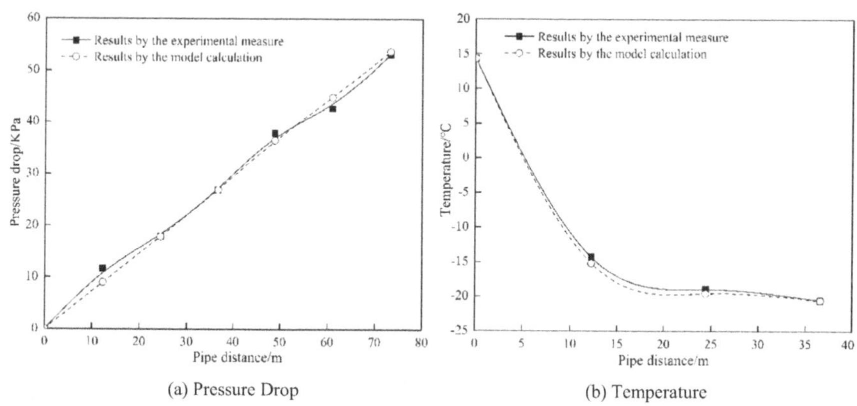

Fig. 5.33 Result of the model calculation and experimental measure

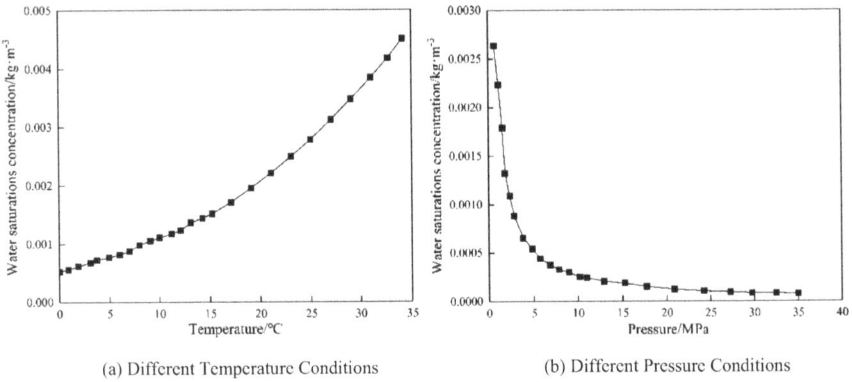

Fig. 5.34 Water saturations concentration of gas at different temperature and pressure

Overall, for the saturated gas flow, the temperature and pressure decreasing along the pipeline would cause the water saturations to be lower and the free water begins to be condensed, promoting the hydrate formation and deposition to the pipewall.

5.5 Prediction for Hydrate Blockage During Deepwater Gas Well Control

In annular-mist flow, the flow patterns of gas-liquid-solid three-phase flow, gas-solid two-phase flow, and water-saturated gas single-phase flow could coexist successfully side by side in the pipeline during deepwater gas well control. In this chapter, the hydrate blockage rules in different flow patterns are calculated by the hydrate deposition and blockage model, and their influencing factors are discussed, further revealing the influence of hydrate blockage on wellhead backpressure during well control and providing a certain reference for the hydrate prediction and prevention on-site [50].

5.5.1 Basic Parameters in Case

The calculation case of well #1 is located in the deepwater area and its basic data is shown in Table 5.4. That the flow pattern belonging to the gas-liquid two-phase flow during well control is calculated by the multiphase flow equation. The liquid holdup along with the well depth at a different time of well control is shown in Fig. 5.35, illustrating the liquid holdup in well annulus and choke lines is low in initial period of well control. Herein, the flow pattern in well annulus and choke lines is annular-mist flow.

Table 5.4 Basic parameters in well #1

Parameters	Value	Gas components	Molar fraction/%
Water depth/m	2000	CH_4	92.55
Well depth/m	3500	C_2H_6	4.59
The inner diameter of choke line/mm	88.9	C_3H_8	1.15
The outer diameter of drill string/mm	216.8	i-C_4H_{10}	0.25
Heat conductivity coefficient of seawater/W $(m{\cdot}k)^{-1}$	1.73	n-C_4H_{10}	0.24
Heat conductivity coefficient of steel/W $(m{\cdot}k)^{-1}$	43.2	i-C_5H_{12}	0.12
Specific heat of seawater/J $(kg{\cdot}K)^{-1}$	3890	n-C_5H_{12}	0.08
Seawater density/$kg{\cdot}m^{-3}$	1025	C_{6+}	0.47
Original stratum pressure/MPa	39	CO_2	0.28
Sea surface temperature/°C	23	N_2	0.29
Geothermal gradient/$K{\cdot}m^{-1}$	0.03		
Relative density of natural gas	0.631		
Stratum temperature/°C	48		
Gas flow/10^4 $m^3{\cdot}d^{-1}$	40		
Liquid mass/$m^3{\cdot}d^{-1}$	80		

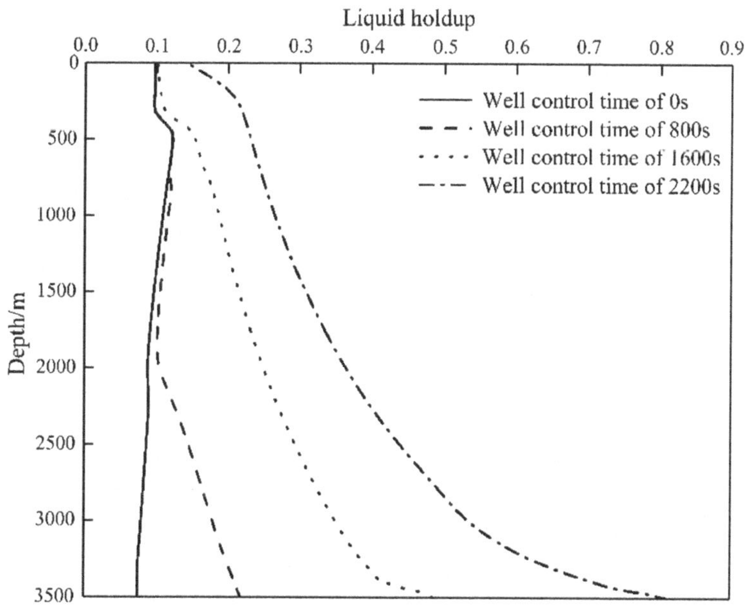

Fig. 5.35 Liquid holdup at different well control time

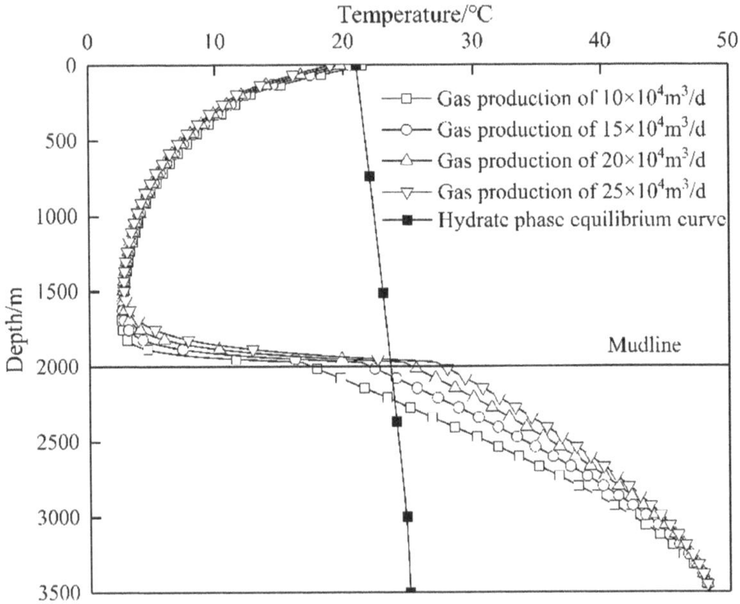

Fig. 5.36 Evolution of hydrate formation area under different gas production

5.5.2 Prediction for Hydrate Formation Area

After the temperature fields in well annulus and choke lines are calculated, by comparing them with the hydrate phase equilibrium curve, the hydrate formation area under different gas production could be obtained in Fig. 5.36, illustrating the hydrate formation area is 0–1942 m containing the entire choke line. Besides, the highest subcooling degree will be found in the choke line near the water depth of 1280 m and it is most likely to accelerate the hydrates formation at such position. On the other hand, with the decrease of gas flow volume, the temperature in the choke line would decrease, rendering the hydrate formation area gradually enlarge. This is because the lower flow volume of gas would provide the more sufficient heat exchange between the fluid in pipe and external environment with low-temperature, leading to the larger temperature drop beyond all doubt, which is easy to meet the temperature conditions that hydrate formation requires.

5.5.3 Laws of Hydrate Deposition and Blockage

Due to the small inner diameter of pipeline and more heat loss between pipe fluid and external environment, the choke line located in the low-temperature seawater is always regarded as a dangerous area in which hydrates blockage is high probability event. After the hydrate formation area in the choke line is determined, hydrate deposition and blockage in choke line during deepwater gas well control could be predicted by solving the hydrate deposition and blockage model. Furthermore, the influence of flow volume of gas and liquid, water depth, pipe diameter on δ_D is analyzed to ultimately reveal the mechanism of hydrate deposition and blockage.

5.5.3.1 Analysis on Basic Rules

As the hydrates are formed and deposited, the flow pattern of gas-liquid-solid three-phase flow, gas-solid two-phase flow, and water-saturated gas flow would be observed successively in choke line, which has different mechanisms and rules of hydrates deposition.

The flow volume of fluid changing along with the choke line is shown in Fig. 5.37. Due to the continuous hydrate formation in the choke line, the free water is gradually consumed so that the flow volume of fluid continues to reduce along the choke line. And, the gas-liquid-solid three-phase flow is here to stay. Nevertheless, at 1257 m, all the free water is converted and generated to the hydrate particles and the flow pattern will be transformed into the gas-solid two-phase flow.

The hydrate particles distribution in the choke line is shown in Fig. 5.38, illustrating the concentration of hydrate particles in the gas phase gradually increases and reaches the "extreme point" in one position, which is the result of the balance

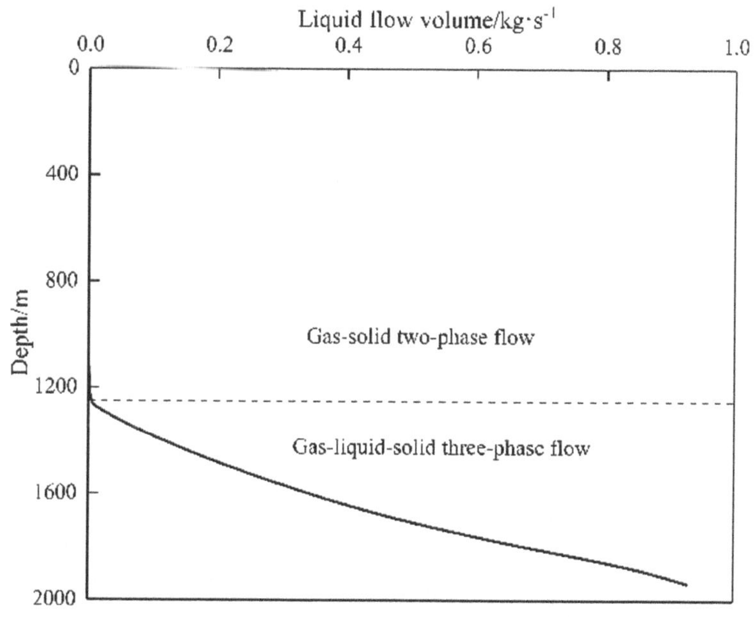

Fig. 5.37 Changes of liquid flow volume

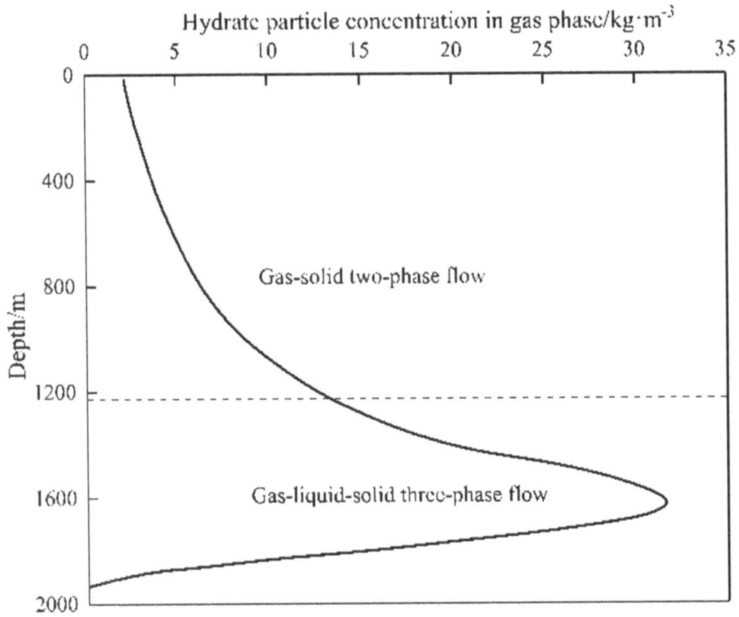

Fig. 5.38 Changes of hydrate particles concentration

between hydrate formation and deposition in gas-liquid-solid three-phase flow. To be more specific, the amount of hydrate formation before the "extreme point" is greater than the amount of hydrate deposition on pipewall, and their opposite relationship would be found after the "extreme point". For gas-solid two-phase flow, the amount of hydrate particles in the gas phase similarly continues to decrease along the pipeline owing to their constant deposition. If the pipeline is long enough, the whole hydrate particles will be attached and deposited on pipewall and no more hydrate particles could be found in the gas phase, making the flow pattern in choke line shift into the water-saturated gas systems.

The changes of δ_D in the choke line are shown in Fig. 5.39, showing δ_D is very unevenly distributed at different pipe positions due to the different rates of hydrate formation and deposition which is caused by different temperatures and pressure. Furthermore, the growth rate of the hydrate layer on pipewall is closely related to the flow pattern and the fastest growth rate would be found in the gas-liquid-solid three-phase flow. Similar to the concentration of hydrate particles in the gas phase, δ_D also presents the "extreme point" and have the same reason why the maximum value appears. So, the position corresponding to the "extreme point" has the most probability for hydrate blockage. It is also found that the growth rate of the hydrate layer in gas-solid two-phase flow is significantly smaller than that in gas-liquid-solid three-phase flow, because the free water could be very beneficial to accelerate the deposition of hydrate particles in the gas phase.

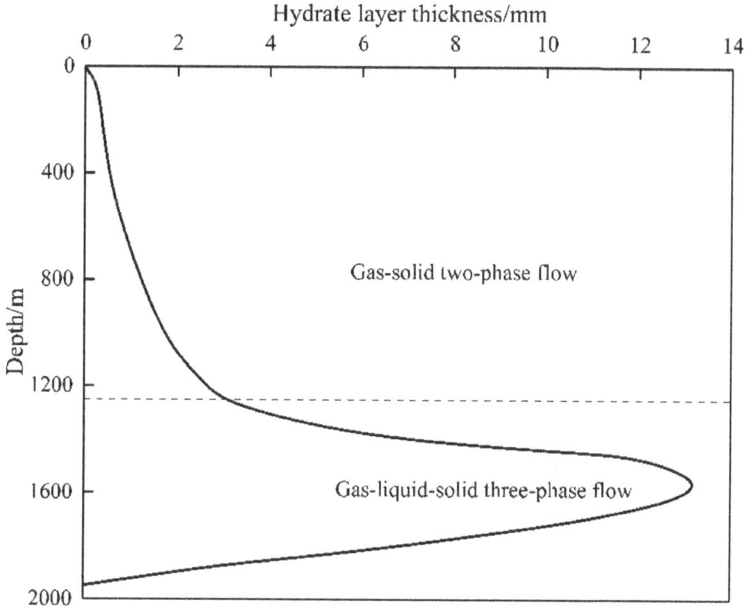

Fig. 5.39 Changes in δ_D

5.5.3.2 Influence of Time

The distribution of δ_D along the pipeline at a different time is shown in Fig. 5.40, which shows that the δ_D in choke line gradually increases with time. The hydrate layer is very unevenly distributed in the whole pipe and its non-uniformity becomes more obvious with time due to the constant deposition of hydrate particles. Besides, the hydrate deposition rate is larger and the hydrate layer will accelerate its thickening process within 1413–1812 m, where has higher hydrate blockage risk.

The interface of gas-liquid-solid three-phase flow and gas-solid two-phase flow is increasingly keeping away from the inlet of choke line (i.e., the position of mudline) as time goes on. It will mainly be because the gradual thickening of the hydrate layer means the reduction of effective inner diameter of the pipe, causing the E and hydrate formation to be smaller under the constant flow volume. Hence, the free water begins to gain larger motion distance as free water is fully consumed and converted to the hydrates.

5.5.3.3 Influence of Gas Flow Volume

The δ_D distribution along the pipeline in different gas flow volumes is shown in Fig. 5.41, illustrating at different well depths, the variation of gas flow volume would exert different influences on the growth of the hydrate layer. In the lower part of

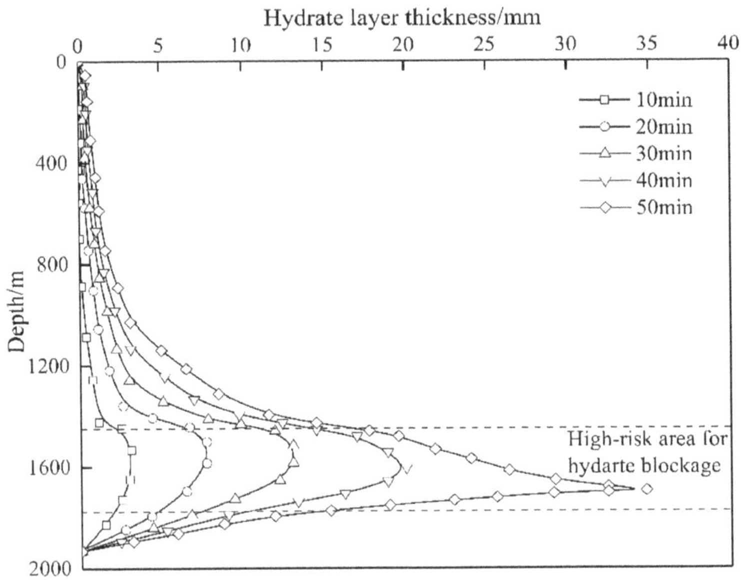

Fig. 5.40 Distribution of δ_D over time

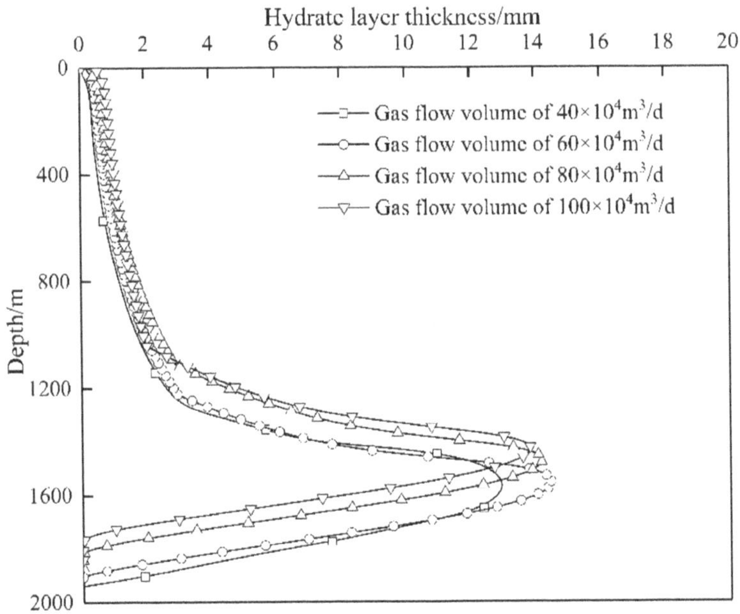

Fig. 5.41 Distribution of δ_D in different gas flow volume

the choke line, δ_D decreases with gas flow volume, while there is a proportionality between δ_D and gas flow volume in the upper part of the choke line.

For the lower part of the choke line, the larger gas flow volume brings about the higher value of E, which is conducive to hydrate formation and deposition. But, there's another side to all of this, the larger gas flow volume also means the insufficient heat exchange between the fluid in pipeline and external environment with low-temperature, named the smaller subcooling degree, which is detrimental to hydrate formation and deposition. Herein, the inhibitory effect of insufficient heat exchange on hydrate formation and deposition remarkably exceeds the promoting effects of the higher E on that. Therefore, δ_D would decrease with the gas flow volume in the lower part of the choke line. Otherwise, under the smaller gas flow volume, the promoting effect of sufficient heat exchange on hydrate formation and deposition is far greater than the inhibitory effects of the lower E on that, resulting in that δ_D starts to grow larger.

In the upper part of the choke line, the temperature of seawater starts moving up gradually, causing that the affecting degree of heat exchange between pipeline fluid and external environment on hydrate formation becomes weaken. Herein, hydrate formation and deposition would be primarily controlled and dominated by the E. Hence, δ_D decreases with the reduction of gas flow volume in the upper part of the choke line.

5.5.3.4 Influence of Liquid Flow Volume

The distribution of δ_D along the pipeline in different volume of liquid flow is shown in Fig. 5.42, illustrating δ_D at different depths all increases with liquid flow volume. This is because the larger liquid flow volume means the gas-liquid contact area and E will increase, which will further accelerate the hydrate formation and deposition. Meanwhile, the interface between gas-liquid-solid three-phase flow and gas-solid two-phase flow will keep away from the inlet of choke line (i.e. mudline position). This is mainly because in the larger liquid flow volume, the free-water is suitable for or reaching long distances until they are fully consumed and converted to the hydrates.

5.5.3.5 Influence of Water Depths

The distribution of δ_D along the pipeline in different water depths is shown in Fig. 5.43, illustrating the water depths mainly influence where hydrate deposition and blockage exactly happens in choke line, but it is less important to the risk levels of hydrate blockage, which is caused by the little difference of seawater temperature outside the choke line. Additionally, the deeper water depth, the smaller δ_D changes near the sea level become. For instance, under the condition of well depth is 3500 m, δ_D shows almost no change in water depth within 0–250 m. This is because the concentration of hydrate particles along the pipeline gradually reduces as the

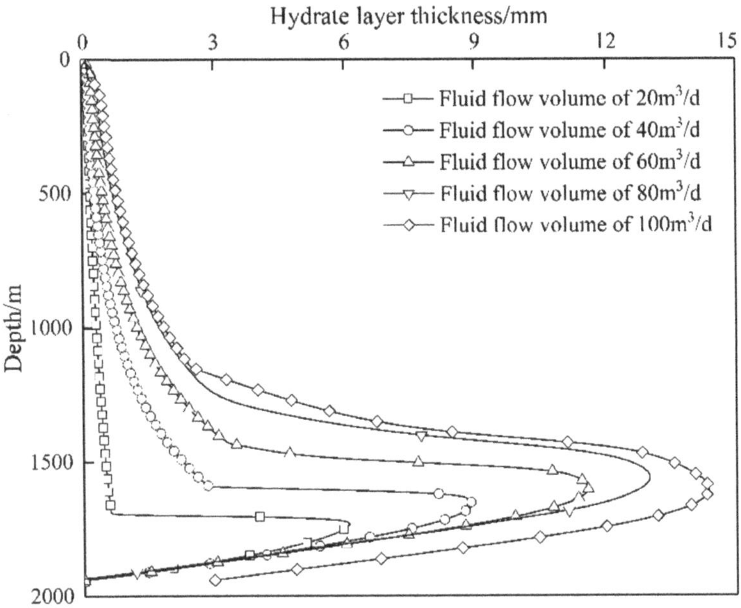

Fig. 5.42 δ_D distribution in different liquid flow volume

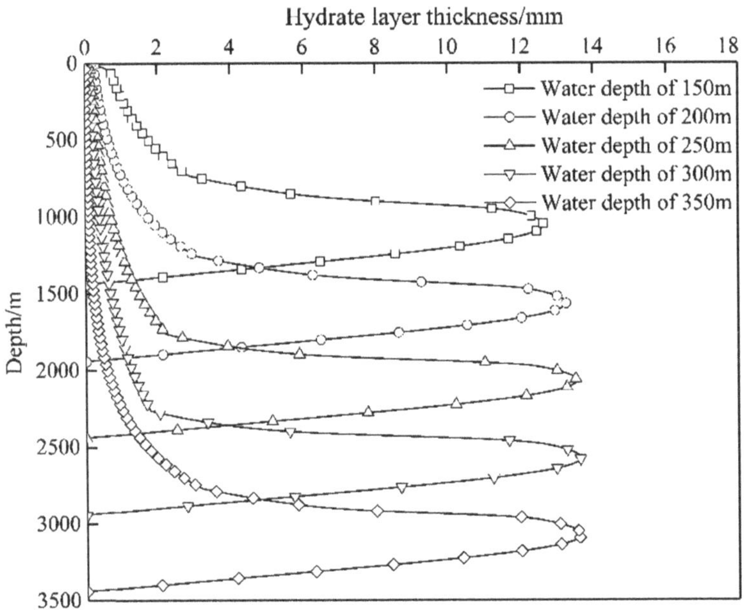

Fig. 5.43 δ_D distribution in different water depths

hydrate particles are deposited and attached to the pipewall, shifting from the gas-solid two-phase flow to the water-saturated gas flow eventually. In this flow pattern, lack of free water severely limits the growth of the hydrate layer, so that it often takes dozens of hours to observe the evident hydrates formation in the experiment. Because of the short experiment time, the value of δ_D in water-saturated gas flow is almost zero.

5.5.3.6 Influence of Pipe Diameter

The δ_D distribution along the pipeline in different diameter of choke line is shown in Fig. 5.44, showing the growth of hydrate layer is accelerated and the hydrate blockage risks are growing as the pipe diameter reduces, which is mainly caused by two aspects: liquid film atomization and the accumulated amount that hydrate blockage requires. Concretely, the size of the cross-sectional area depends on the pipe diameter. The smaller pipe diameter, the smaller cross-sectional area and the greater apparent velocity of gas are. Herein, liquid film atomization would occur under the action of a bigger gas shear force, leading to a higher rate of hydrate formation and deposition. On the other side, compared with the larger pipe diameter, the accumulated amount that hydrate blockage requires is less in the pipe with smaller diameter, resulting in that the smaller pipeline diameter tends to form the hydrate layer and becomes more prone to the high-risk hydrate blockage.

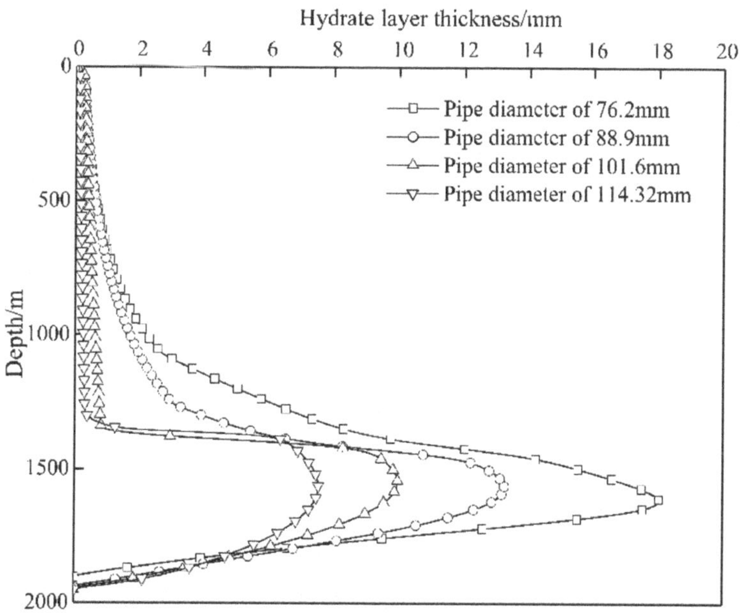

Fig. 5.44 δ_D distribution in different diameters of pipe

5.5.4 *Influence of Hydrate Deposition on Wellhead Back Pressure*

The changes of wellhead backpressure in different gas production are shown in Fig. 5.45, showing that under the high gas production, the wellhead backpressure decreases marginally first and then evolves toward a steady-going value, finally suffer a cliff-like drop. δ_D directly influences the frictional pressure drop of annular-mist flow in the pipeline [51], so the growth rate of the hydrate layer determines the developing trend of frictional pressure drop which will slow a little or sharply. Besides, the greater gas production, the faster wellhead backdrops because of the higher frictional resistance.

In on-site conditions of deepwater well control, the wellhead backpressure, as an important parameter that can be directly monitored, could be effectively used to predict the risks intensity and severity for hydrate deposition and blockage in choke line, which could bright some light to the operation safety during deepwater well control on-site.

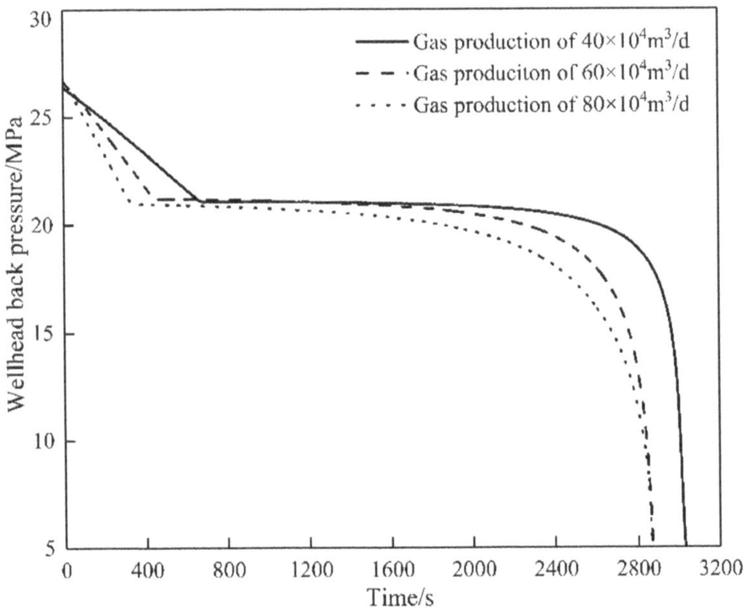

Fig. 5.45 Changes of wellhead back pressure in different gas production

5.6 Prediction for Hydrate Blockage During Deepwater Gas Well Test

5.6.1 Hydrate Blockage Free Window (HBFW)

As the hydrate layer on pipewall continuously thickens, the cross-sectional area of the pipeline would decrease and the surface roughness of pipe would increase, causing the frictional pressure drop in the pipeline to increase remarkably.

Based on the multiphase flow model in wellbore and hydrate blockage model, the relation between frictional pressure drop and dimensionless δ_D could be obtained shown in Fig. 5.46 [52], illustrating the frictional pressure drop increases significantly and throttling effect appears evidently when nondimensional δ_D reaches 0.45.0.55 due to the hydrate deposition. More generally, such a phenomenon will also appear in different gas flow rates, temperatures, pressures, the void fraction (less than 10%). Herein, it's inevitable that 0.5 is ideally suitable to be regarded as the nondimensional critical hydrate layer thickness (δ_{Dc}) and as a good index or sign when hydrate blockage happens. When $\delta_D < \delta_{Dc}$, the frictional pressure drop is so small that the stratum fluid can flow smoothly to the ground and no prevention and management measures are needed. Otherwise, when $\delta_D > \delta_{Dc}$, the frictional pressure drop continues to be enlarged so that the hydrate blockage risks increase and normal operations cannot be usually performed, then hydrate management measures need to be taken immediately.

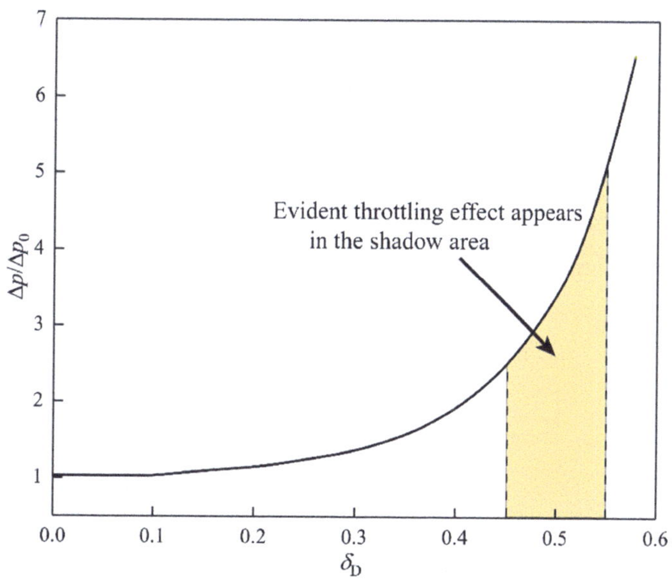

Fig. 5.46 Relation of frictional pressure drop and nondimensional δ_D

Wang et al. and Sun et al. [2, 25, 52–57] firstly proposed the concept of hydrate blockage free window (HBFW), and successfully carried out the quantitative prediction and effective control on the hydrate blockage. Herein, the lasting time when nondimensional δ_D at the initial moment reach the value of nondimensional δ_{Dc} is defined as the HBFW, which aims to analyze the hydrate deposition dynamics in the pipe column. Determining the HBFW and concrete blockage position could provide more guidance for hydrate prevention and management in the deepwater gas well test.

5.6.2 Model Solution Steps

The interrelationship of hydrate formation and deposition, hydrate layer growth, temperature and pressure distribution in the system leads to the strong nonlinearity of the hydrate blockage model. Its numerical solution in different flow patterns is performed with the finite difference method shown in Fig. 5.47.

(1) Set small enough interval dt (dt = 1 s) and divide the pipe column/pipeline into multiple-unit control bodies with a length of dl (dl = 1 m) shown in Fig. 5.48. In

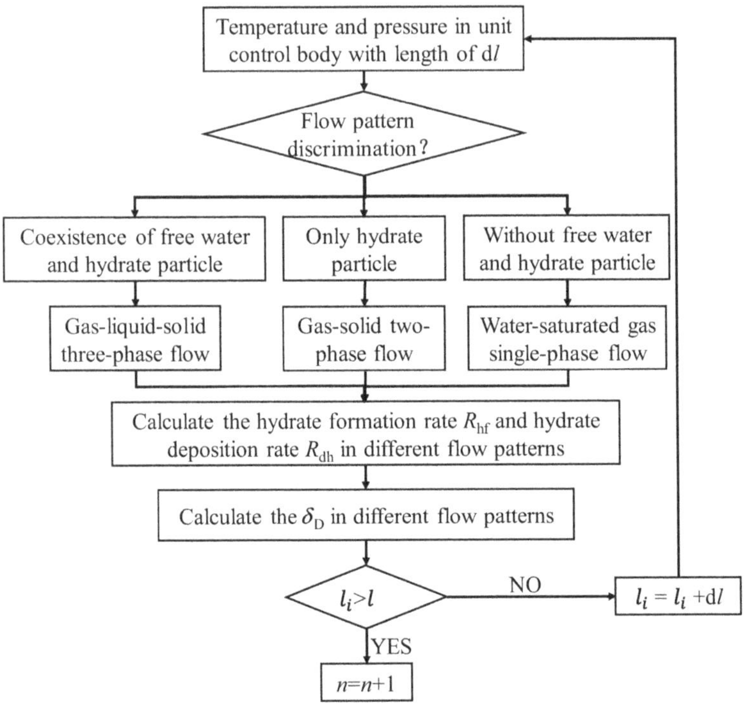

Fig. 5.47 Flow chart for solving the hydrate deposition and blockage in different flow patterns

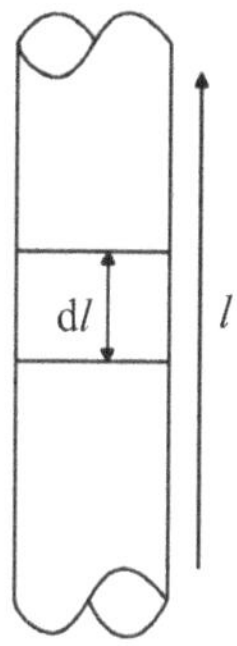

Fig. 5.48 Diagram of unit control bodies division

dl, assume that: ① The temperature and pressure distribute uniformly, so the hydrate formation rate at different positions remains constant; ② The droplets in the gas phase and hydrate particles distribute uniformly, so the hydrate deposition rate at different positions is the same; ③ The hydrate density remains constant and the porosity of hydrate layer on pipewall is neglected; ④ The heat transfer only occurs at the radial direction of pipe column; ⑤ The hydrate layer detaching and removing from the pipewall is neglected.

(2) Obtain the initial temperature and pressure in first dl, then calculate the temperature and pressure in the next dl according to energy conservation. Notice the temperature and pressure at the inlet of unit control body $i + 1$ is equal to the outlet value of unit control body i.

(3) According to the changes of free water and hydrate particle amount in the system, the flow type at different pipe column positions is simply identified: ① The initial flow type is gas-liquid-solid three-phase flow in pipe column when gas, free water, and hydrate particles coexist; ② Only the gas and hydrate particles coexist in pipe column when the free water is fully consumed and converted to the hydrates, presenting the gas-solid two-phase flow; ③ The water-saturated gas system appears when all the hydrate particles are deposited and attached on pipewall.

(4) Calculate the rate of hydrate formation and deposition in each flow pattern, then obtain the value of δ_D at different pipeline positions.

(5) Repeat the above steps to calculate the hydrate formation and deposition rate, δ_D in next dl when the length l_i is smaller than the pipe column length l_s; otherwise, calculate δ_D at different positions the next moment until the required time.

5.6.3 *Quantitative Prediction for Hydrate Blockage*

Based on the wellbore multiphase flow model, the hydrate blockage in wellbore could be quantitatively predicted by analyzing the characteristics of hydrate formation, deposition and blockage, which could be solved and determined by the following steps [30]:

(1) Calculate the temperature and pressure distribution in the wellbore to determine the hydrate formation area

Firstly, the temperature and pressure distribution in the wellbore is determined based on the multiphase flow model established in Chap. 4, and the hydrate formation area could be determined in view of the hydrate phase equilibrium theory. The diagram of quantitative prediction for the hydrate blockage during the deepwater gas well test is shown in Fig. 5.49 [51], where the left side shows the determination of the hydrate formation area and the right side represents the quantitative prediction result of hydrate blockage.

(2) Calculate the rate of hydrate formation and deposition

After the hydrate formation area is determined, the hydrate formation and deposition rate at different wellbore positions could be calculated according to the hydrate formation rate model described in Chap. 2 and the hydrate deposition model in this chapter.

(3) Predict the hydrate blockage dynamically

The value of nondimensional δ_D at different positions could be obtained based on hydrate deposition rate, and the HBFW could be determined by simulating the developing and evolution process of hydrate blockage, achieving the dynamic prediction for hydrate blockage. Apparently, the quantitative prediction results of hydrate blockage, shown in Fig. 5.49, briefly introduces and clarifies the HBFW, hydrate

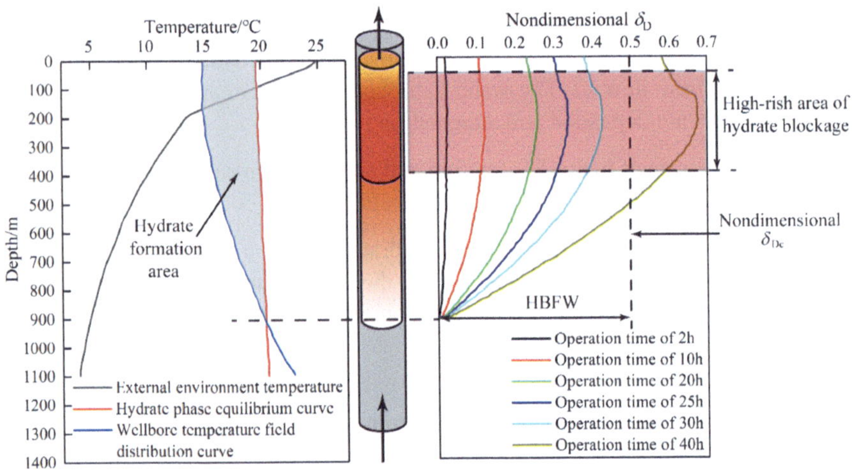

Fig. 5.49 Diagram of quantitative prediction for the hydrate blockage

blockage position and its dynamic evolution, further providing the basis for the scheme design of hydrate management.

5.6.4 *Case Analysis for Predicting the Hydrate Blockage*

In this section, the quantitative prediction and early monitoring of hydrate blockage, new hydrate prevention and management methods based on HBFW are described in view of the multiphase flow model and hydrate blockage model, then the influence of gas flow volume and water depth on hydrate blockage is analyzed.

5.6.4.1 Basic Parameters in Deepwater Gas Well

In deepwater gas well, the conditions of high-pressure and low-temperature near the sea bed are of great helpful to the hydrate formation. After drilling, the gas well test is first required to clarify the reservoir properties and potential gas production rate, and to provide the basic data related to formulate the subsequent development scheme. By now, the deepwater gas well #3 in a block of one sea area is taken for the case analysis. Its basic parameters are shown in Table 5.5 and the diagram of fluid flow and hydrate deposition in gas well are illustrated in Fig. 5.50.

Table 5.5 Basic parameters of deepwater gas well #3 in a block of one sea area

Basic parameters	Value	Basic parameters	Value
Wellbore type	Straight well	Flow type in the wellbore	Ring mist flow
Heat conductivity coefficient of steel/W $(m \cdot k)^{-1}$	43.2	Natural gas relative density	0.554
Heat conductivity coefficient of stratum/W $(m \cdot k)^{-1}$	2.2	Well depth/m	3800
Heat conductivity coefficient of seawater/W $(m \cdot k)^{-1}$	1.73	Water depth/m	1500
Heat conductivity coefficient of cement/W $(m \cdot k)^{-1}$	0.35	Mean depth of test position/m	3800
Specific heat of stratum rock J/(kg·K)	830	Sea surface temperature/°C	25
Seawater specific heat J/(kg·K)	3890	Geothermal gradient/°C·m^{-1}	0.0192
Seawater density/kg·m^{-3}	1025	Bottom hole pressure/MPa	28
Stratum rock density/kg·m^{-3}	2640	Wellhead pressure/MPa	18
Gas output/$(m^3 \cdot d^{-1})$	45×10^4	Liquid-phase volume fraction in the pipe column /%	3
Water output/$(m^3 \cdot d^{-1})$	15		

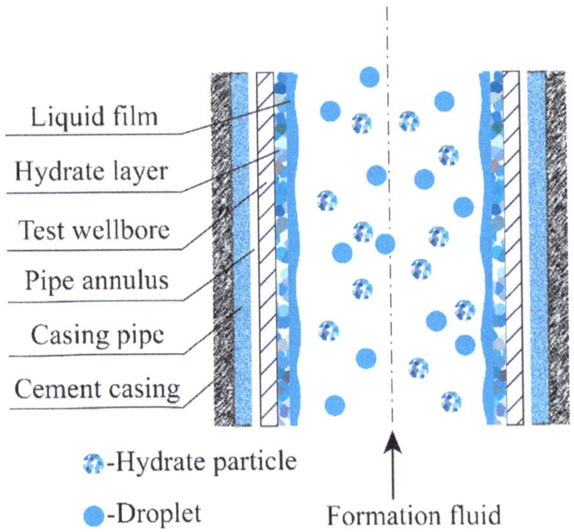

Fig. 5.50 Fluid flow and hydrate deposition in deepwater gas well #3

5.6.4.2 Dynamic Evolution for Hydrate Blockage in Test Pipe Column

According to the parameters shown in Table 5.5 and the multiphase flow model, the wellbore temperature field is obtained. By comparing the temperature field curve with the hydrate phase equilibrium condition, the hydrate formation area is shown in Fig. 5.51 [30].

The ambient temperature near the sea bed is low, but the fluid temperature in well is higher than the phase equilibrium temperature for hydrate formation (T_{eq}), so no hydrates are converted near the sea bed. Afterwards, as the upward migration of fluid, the heat exchange is extremely facilitated and the heat is lost very greatly, causing that the fluid temperature at 900 m is lower than the T_{eq}. That is to say, the fluid at this position has entered the hydrate formation area extending to the platform wellhead, so the hydrate formation area is 0–900 m.

Based on the hydrate formation area and hydrate blockage model, the blockage characteristics in the testing pipe could be obtained. The variations of δ_D with time at different wellbore positions are shown in Fig. 5.52. As can be seen from the figure, the more time passes the higher value of δ_D and the less effective diameter of the pipe it gets, but it takes different time for δ_D to reach the δ_{Dc}. The essential reason for this phenomenon is that the temperature and pressure of fluid, subcooling degree are different at various wellbore positions, rendering the hydrate formation and deposition rate different. Also, at the beginning of well testing, there is little change in δ_D and pipeline diameter, causing the gas production to proceed smoothly. Nevertheless, the depth range of 50–400 m, displaying the higher growth rate of the hydrate layer and blockage risks, is consequently defined as the hydrate blockage high-risk area. Notably, the growth rate of the hydrate layer at 150 m reaches the

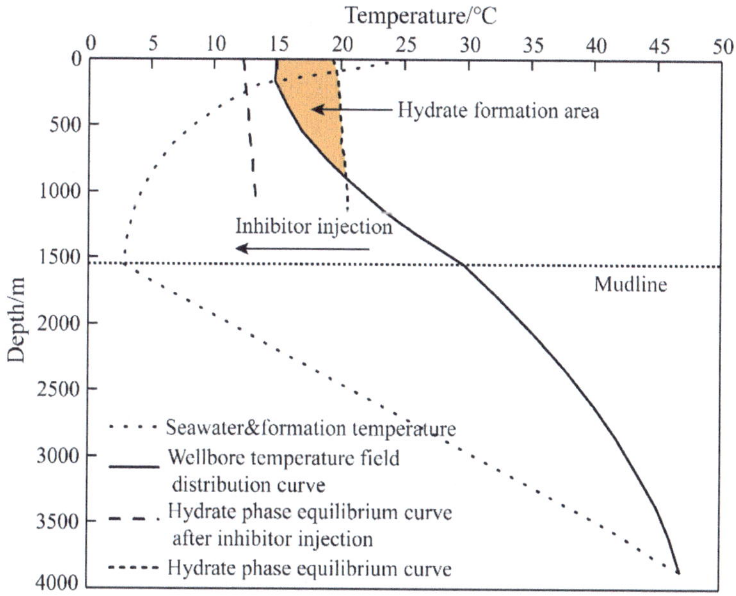

Fig. 5.51 Hydrate formation area of test pipe column

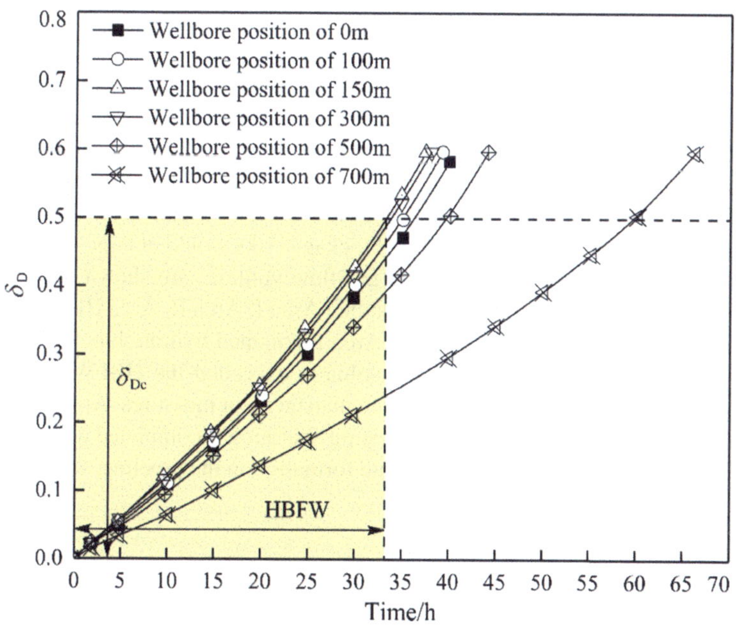

Fig. 5.52 Change of δ_D with time at different wellbore positions

maximum. Therefore, at this position, the lasting time when hydrate layer achieves the δ_{Dc} is conventionally defined as the HBFW, which is 33.3 h in this case.

In light of the above consideration, two criteria for identifying the hydrate blockage are summarized as follows:

(1) Judge whether the given position of wellbore is located at the hydrate formation area
This criterion is mainly used to judge whether hydrates are converted and formed at the given position of the wellbore during well testing. If it is, the hydrate blockage shall be judged based on the second criterion.
(2) Compare the HBFW with testing time
If the HBFW is longer than the testing time, a small amount of hydrates would be formed and generated in the pipe column during well testing, but exert a negligible influence on the testing operation, so the normal testing could be conducted as planned. Otherwise, hydrate blockage would occur if any corresponding hydrate management measures are not taken timely and effectively. But overall, hydrates are allowed to be formed continuously during well testing until the testing time reaches the HBFW, and the related management must be carried out below HBFW for the safe construction.

5.6.4.3 Influence of Gas Flow Volume on Hydrate Blockage

Keeping other conditions unchanged, changing the gas flow volume could obtain the changes of δ_D with time at 1500 m shown in Fig. 5.53. It illustrates that the HBFW is shortened and the hydrate blockage risks are increased accordingly with the gas production. This is because the heat exchange between well fluid and ambient temperature is much more sufficient at the lower gas production, amplifying the subcooling degree and promoting the hydrates formation and deposition. Hence, both hydrate formation area and hydrate blockage risks would enlarge accordingly.

The variations of HBFW in different gas flow volumes are shown in Fig. 5.54, which could be divided into three areas named Area I, Area II, Area III respectively according to the forementioned criteria. Area I is named hydrate blockage area, in which the testing operation time has absolutely exceeded the HBFW, so hydrate blockage will be found without any hydrate management measures. Area II is called the hydrate formation area, in which the testing operation is completed within HBFW and there is only a small amount of hydrate formation in the pipeline, so the hydrate

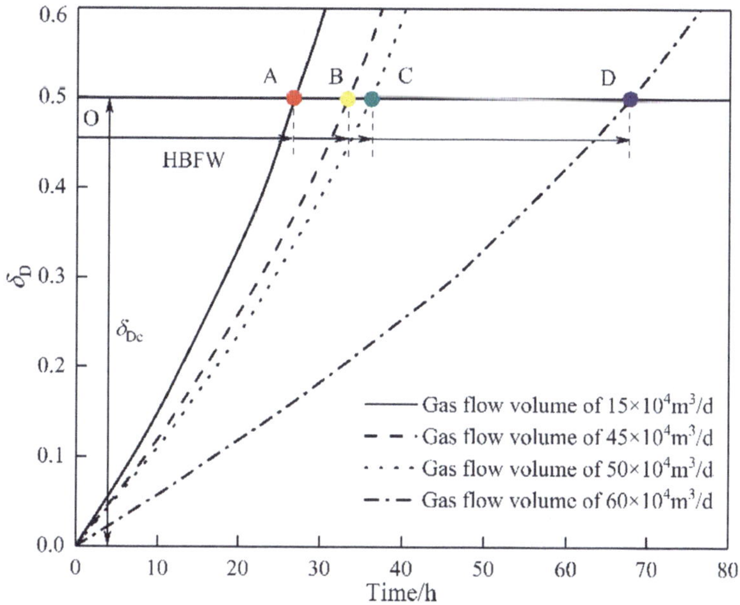

Fig. 5.53 Changes of δ_D with time at different flow volume of gas (water depth of 1500 m)

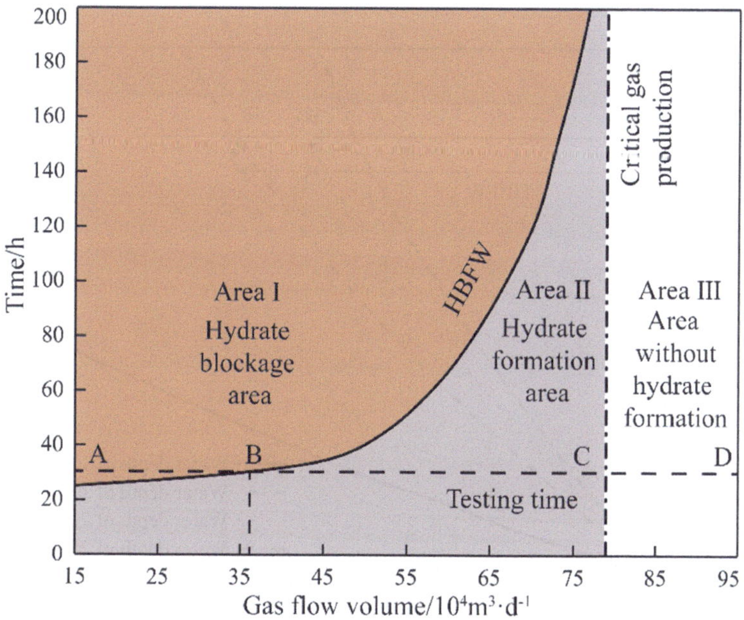

Fig. 5.54 HBFW in different gas flow volume (water depth of 1500 m)

blockage will not happen and the testing could be conducted normally as planned. Area III is named area without hydrate formation. When the gas production is greater than one critical gas production, the whole process of well testing is located outside Area II and hydrate formation, not to mention the hydrate blockage risks, is not observed in such area.

Suppose the well testing time is 30 h. The gas flow volume first enters into the range of A-B, and the well is located in the Area I. Afterwards, the well would depart from Area I and begin to get into the Area II until the gas flow volume reaches the critical value. Finally, no hydrates are generated in Area III. Apparently, critical gas production has a more important guiding significance for the subsequent design of production parameters.

5.6.4.4 Influence of Water Depth on Hydrate Blockage

Keeping other conditions unchanged, changing the water depth could obtain the variations of δ_D with time at wellbore gas production of 45×10^4 m^3/d, which is shown in Fig. 5.55 [51], illustrating the HBFW is shortened and the hydrate blockage risks are increased accordingly as the water depth decreases. This is because the larger water depth provides the lower bottom hole temperature in the same gas production, so both the subcooling degree and the hydrate formation area would be enlarged.

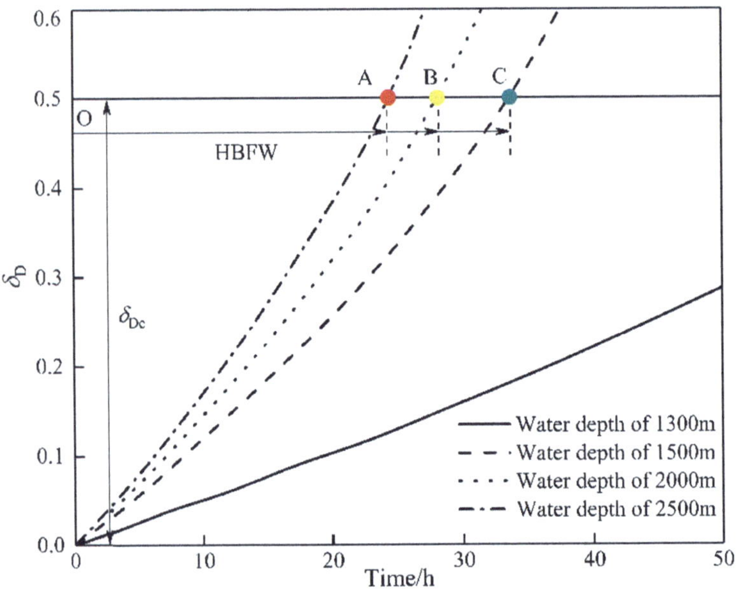

Fig. 5.55 Changes of δ_D with time at gas production of 45×10^4 m^3/d

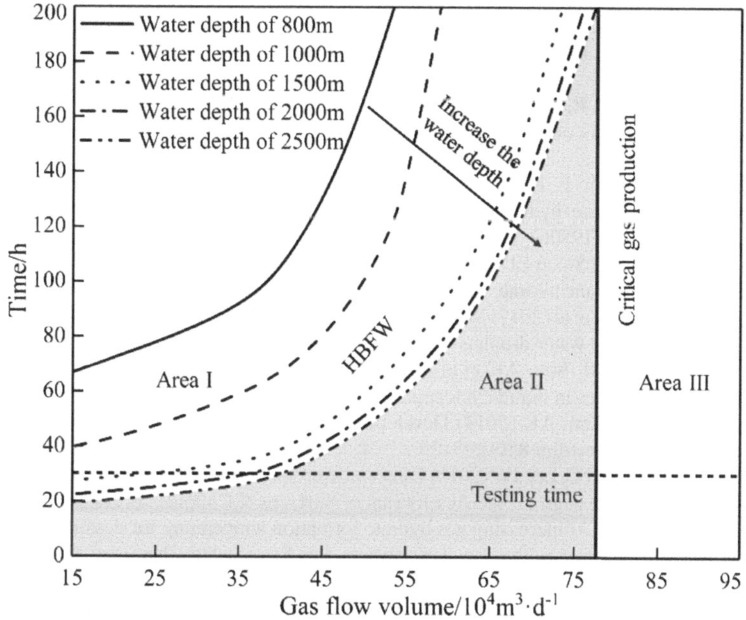

Fig. 5.56 HBFW in different water depths

The HBFW variations at different water depths are shown in Fig. 5.56. As can be seen from the figure, the HBFW would gradually shift to the lower-right corner with the water depth, while its shifting rate would be getting smaller. Consequently, Area I becomes larger, whereas Area II becomes smaller, representing the higher hydrate blockage risks. Herein, it is necessary to take relevant management measures immediately for plug removal.

References

1. Wang Z, Zhang J, Chen L et al (2017) Modeling of hydrate layer growth in horizontal gas-dominated pipelines with free water. J Nat Gas Sci Eng 50:364–373
2. Zhang J, Wang Z, Sun B et al (2019) An integrated prediction model of hydrate blockage formation in deep-water gas wells. Int J Heat Mass Transf 140:187–202
3. Yang S, Kleehammer DM, Huo Z et al (2003) Micromechanical measurements of hydrate particle attractive forces//15th Symp on Thermophysical Properties, Boulder, USA
4. Herri FAGF (2006) Rheology of methane hydrate slurries during their crystallization in a water indodecane emulsion under flowing. Chemical Engineering Science (61):505–515
5. Camargo R, Palermo T, Sinquin A et al (2010) Rheological characterization of hydrate suspensions in oil dominated systems. Ann N Y Acad Sci 912(1):906–916
6. Zhu C, Li YX, Wang WC (2012) Static liquid bridge force between hydrate particles. Oil-Gas Field Surface Engineering 31(10):26–28
7. Taylor C J, Dieker L E, Miller K T et al (2007) Micromechanical adhesion force measurements between tetrahydrofuran hydrate particles. J Colloid Interface Sci 306(2):255–261

8. Dieker LE, Aman ZM, George NC et al (2009) Micromechanical adhesion force measurements between hydrate particles in hydrocarbon oils and their modifications. Energy Fuels 23(12):5966–5971
9. Aman ZM, Dieker LE, Aspenes G et al (2010) Influence of model oil with surfactants and amphiphilic polymers on cyclopentane hydrate adhesion forces. Energy Fuels 24(10):5441–5445
10. Aman Z M, Brown E P, E Dendy S et al (2011) Interfacial mechanisms governing cyclopentane clathrate hydrate adhesion/cohesion. Physical Chemistry Chemical Physics, 2011,13(44):19796–19806
11. Aman ZM, Joshi SE, Sloan ED et al (2012) Micromechanical cohesion force measurements to determine cyclopentane hydrate interfacial properties. J Colloid Interface Sci 376(1):283–288
12. Liu C, Li Y, Wang W et al (2017) Modeling the micromechanical interactions between clathrate hydrate particles and water droplets with reducing liquid volume. Chem Eng Sci 163:44–55
13. Maeda N, Aman ZM, Sum AK et al (2013) Measurements of cohesion hysteresis between cyclopentane hydrates in liquid cyclopentane. Energy Fuels 27(9):5168–5174
14. Lee BR, Koh CA, Sum AK (2014) Development of a high pressure micromechanical force apparatus. Rev Sci Instrum 85(9):95120
15. Song JHK, Alexander C, Lee JW (2010) Direct measurements of contact force between clathrate hydrates and water. Langmuir the Acs Journal of Surfaces & Colloids 26(12):9187–9190
16. Cha JH, Seol Y (2013) Increasing gas hydrate formation temperature for desalination of high salinity produced water with secondary guests. Acs Sustainable Chemistry & Engineering 1(10):1218–1224
17. Liu HH, Li YX, Wang WC et al (2013) Orthogonal experiment research on factors affecting hydrate accumulation. Oil & Gas Storage and Transportation 32(11):1232–1236
18. Liu CW (2014) Multiphase flow of water-in-oil emulsion by considering hydrate transition. China University of Petroleum (East China)
19. Liu C, Li M, Chen L et al (2017) Experimental investigation on the interaction forces between clathrate hydrate particles in the presence of a water bridge. Energy Fuels 31(5):4981–4988
20. Liu C, Li M, Liu C et al (2016) Micromechanical interactions between clathrate hydrate particles and water droplets: experiment and modeling. Energy Fuels 30(8):6240–6248
21. Yang S (2003) Simulation of hydrate agglomeration by discrete element method. In: 15th Symposium on Thermophysical Properties, Boulder, USA
22. Aspenes G, Dieker LE, Aman ZM et al (2010) Adhesion force between cyclopentane hydrates and solid surface materials. J Colloid Interface Sci 343(2):529–536
23. Wang Z, Yu J, Zhang J et al (2018) Improved thermal model considering hydrate formation and deposition in gas-dominated systems with free water. Fuel 236:870–879
24. Lorenzo MD, Aman ZM, Kozielski K et al (2014) Underinhibited hydrate formation and transport investigated using a single-pass gas-dominant flowloop. Energy Fuels 28(11):7274–7284
25. Wang ZY, Zhao Y, Sun BJ et al (2016) Heat transfer model for annular-mist flow and its application in hydrate formation risk analysis during deepwater gas well testing. Journal of Hydrodynamics 31(1):20–27
26. Jassim E I, Abdi M A, Muzychka Y (2008) A CFD-based model to locate flow-restriction induced hydrate deposition in pipelines. Offshore Technology Conference
27. Jassim E, Abdi MA, Muzychka Y (2010) A new approach to investigate hydrate deposition in gas-dominated flowlines. Journal of Natural Gas Science & Engineering 2(4):163–177
28. Wang Z, Zhang J, Sun B et al (2017) A new hydrate deposition prediction model for gas-dominated systems with free water. Chem Eng Sci 163(145):154
29. Schadel SA, Hanratty TJ (1989) Interpretation of atomization rates of the liquid film in gas-liquid annular-mist flow. Int J Multiph Flow 15(6):893–900
30. Wang ZY, Zhao Y, Sun B et al (2016) Modeling of hydrate blockage in gas-dominated systems. Energy Fuels 30(6):4653–4666

31. Bertodano MALD, Jan CS, Beus SG (1996) Annular-mist flow entrainment rate experiment in a small vertical pipe. Nucl Eng Des 178(1):61–70
32. Assad A, Jan C, Bertodano MLD et al (1998) Scaled entrainment measurements in ripple-annular-mist flow in a small tube. Nucl Eng Des 184(2–3):437–447
33. Nicholas JW, Dieker LE, Sloan ED et al (2009) Assessing the feasibility of hydrate deposition on pipeline walls—Adhesion force measurements of clathrate hydrate particles on carbon steel. J Colloid Interface Sci 331(2):322–328
34. Zhang J, Wang Z, Liu S et al (2019) Prediction of hydrate deposition in pipelines to improve gas transportation efficiency and safety. Appl Energy 253:113521
35. Friedlander SK, Johnstone HF (1954) Deposition of suspended particles from turbulent gas strems. Ind Eng Chem 49(7):1151–1156
36. Wells AC, Chamberlain AC (1967) Transport of small particles to vertical surfaces. Br J Appl Phys 18(12):1793–1799
37. Fichman M, Gutfinger C, Pnueli D (1988) A model for turbulent deposition of aerosols. J Aerosol Sci 19(1):123–136
38. Fan F G, Ahmadi G (1993) A sublayer model for turbulent deposition of particles in vertical ducts with smooth and rough surfaces. Journal of Aerosol Science 24(1):45–64
39. Zhang J, Zhou LX (2000) Analytical solutions of the equation of particle motion along trajectories in gas-solid flows. Journal of Combustion Science and Technology 6(3):226–229
40. WOOD N B (1981) A simple method for the calculation of turbulent deposition to smooth and rough surfaces. Journal of Aerosol Science 12(3):275–290
41. Sosnowski TR, Moskal A, Gradoń L (2006) Mechanisms of aerosol particle deposition in the oro-pharynx under non-steady airflow. The Annals of occupational hygiene 51(1):19–25
42. Kvasnak, AhmadiG, Bayer et al (1993) Experimental investigation of dust particle deposition in a turbulent channel flow. Journal of Aerosol Science 24(6):795–815
43. Tian L, Ahmadi G (2007) Particle deposition in turbulent duct flows—comparisons of different model predictions. J Aerosol Sci 38(4):377–397
44. Shams M, Ahmadi G, Rahimzadeh H (2000) A sublayer model for deposition of nano- and micro-particles in turbulent flows. Chem Eng Sci 55(24):6097–6107
45. Fan G, Xu JY (1994) Theoretical model of turbulent particle concentration in gas-solid two-phase flow field. Chinese Journal of Applied Mechanics 4:11–18
46. Lu HL, Zhao B (1999) Numerical simulated calculation of gas-solid two-phase flow in vertical pipe is theoretical method of particle dynamics. Journal of Mechanical Engineering 35(5):75.79
47. Zhang H, Ahmadi G (2000) Aerosol particle transport and deposition in vertical and horizontal turbulent duct flow. Journal of Fluid Mechanics 406:55–80
48. Lingelem M N, Majeed A I, Stange E (2010) Industrial experience in evaluation of hydrate formation, inhibition, and dissociation in pipeline design and operation. Annals of the New York Academy of Sciences 715(1):75–93
49. Rao I, Koh CA, Sloan ED et al (2013) Gas hydrate deposition on a cold surface in water-saturated gas systems. Ind Eng Chem Res 52(18):6262–6269
50. Zhang JB (2018) Hydrate deposition and blockage mechanism in annular-mist flow in deepwater well control. China University of Petroleum (East China)
51. Wang Z, Yang Z, Zhang J et al (2018) Quantitatively assessing hydrate-blockage development during deepwater-gas-well testing. SPE Journal 23(4):1166–1183
52. Wang Z, Yang Z, Zhang J et al (2018) Flow assurance during deepwater gas well testing: Hydrate blockage prediction and prevention. J Petrol Sci Eng 163:211–216
53. Wang ZY, Yang Zhao Y, Sun BJ et al (2018) Features and prevention of gas hydrate blockage in test strings of deep-water gas wells. Nat Gas Ind 38(01):81–88
54. Wang ZY, Yu W, Meng WB et al (2018) Quantitative prediction of gas hydrate blockage in testing tubing for deepwater gas well testing. China Offshore Oil and Gas 30(03):122–131
55. Deng Z, Wang Z, Zhao Y, et al (2017) Flow assurance during gas hydrate production: hydrate regeneration behavior and blockage risk analysis in wellbore//Abu Dhabi International Petroleum Exhibition & Conference. Society of Petroleum Engineers

56. Zhao Y, Wang Z, Yu J et al (2017) Hydrate plug remediation in deepwater well testing: a quick method to assess the plugging position and severity//SPE Annual Technical Conference and Exhibition. Society of Petroleum Engineers
57. Zhao Y, Wwang Z, Zhang J, et al (2016) Flow assurance during deepwater gas well testing: When and where hydrate blockage would occur//SPE Annual Technical Conference and Exhibition. Society of Petroleum Engineers

Chapter 6
Technologies for Hydrate Management in Deepwater Gas Well

Special attention should be paid to hydrate management throughout the entire process of deepwater oil and gas development. The hydrate blockage in deepwater gas well not only affects the normal gas production but its subsequent removal also prolongs the operation time, increases the operation costs and even jeopardizes the personnel and production equipment. The main methods for hydrate prevention and management are divided into two types: chemical inhibitor methods and physical methods. The physical methods mainly include thermal insulation method, thermal stimulation method, depressurization method, a coating method, dehydration method, gravity pipe heating method, and down-hole gas valve throttling method. For the chemical inhibitor methods, they aim to alter the hydrate phase equilibrium or inhibit the nucleation, growth, and aggregation of hydrates. Common chemical inhibitors mainly include thermodynamic inhibitors, kinetic inhibitors, and anti agglomerants, which are simple and effective compared with the physical methods. This chapter mainly introduces hydrate management during the testing of deepwater gas wells.

6.1 Classification of Chemical Inhibitors

6.1.1 Thermodynamic Inhibitors

The hydrate-prevented mechanism of thermodynamic inhibitors is usually displayed in two aspects. Thermodynamic inhibitors could reduce the water activity, change the thermodynamic equilibrium between water and gas molecules, and alter the chemical potential of aqueous solution or hydrates, causing the hydrate phase equilibrium curve to shift to the left. Besides, some active components in inhibitors are directly brought into contact with hydrates to make them unstable. Thermodynamic inhibitors are mainly comprised of alcohol and inorganic salt [1–4]. Alcohols mainly include methanol, ethylene glycol, isopropanol, diethylene glycol and the like. The methanol,

© Science Press and Springer Nature Singapore Pte Ltd. 2020
Z. Wang et al., *Natural Gas Hydrate Management in Deepwater Gas Well*,
https://doi.org/10.1007/978-981-15-6418-5_6

Table 6.1 Physical properties of methanol, ethanediol and triethylene glycol

Type	Methanol	Ethanediol	Triethylene glycol
Molecular formula	CH_3OH	$C_2H_4(OH)_2$	$O(CH_2CH_2OH)_2$
Molecular weight	32.04	62.07	106.1
Density/g·cm^{-3}	0.7915	1.1088	1.1184
Freezing point/°C	−97.8	−13	−8.3
Boiling point/°C	64.7	197.3	245.0
Flash point/°C	15.6	116	123.9
Vapor pressure at 20 °C	92	<1	0.13
Viscosity at 20 °C/Mpa·s	0.6	20.9	35.7
Water solubility	Mutually soluble	Mutually soluble	Mutually soluble
Characteristic	Colorless, inflammable, volatile, medium toxic	Colorless, non-toxic, non-corrosive	Colorless, low-toxic, non-corrosive, sticky

ethylene glycol and diethylene glycol, as widespread used alcohols, have many physical properties as shown in Table 6.1. For the salt substances, they mainly include sodium chloride, calcium chloride, magnesium chloride and lithium chloride and the like. Calcium chloride and sodium chloride are the most commonly used on-site, but their corrosion limits the application frequency.

Utilization of thermodynamic inhibitor has many disadvantages such as high consuming amount, large storage and injection equipment, and environmental pollution. What's more, they work well only at higher concentrations (mass fraction exceeding 6%). At low concentration (1%–5%), hydrates formation would be promoted rather than being inhibited. Therefore, it is necessary to design the injection parameters reasonably and to strive for the maximum inhibitory effect using the smallest volume of inhibitors.

6.1.2 Kinetic Inhibitors

Kinetic inhibitors are mainly adsorbed onto the hydrate crystal and water interface, which reduces the hydrate formation rate, prolongs the induction time of hydrate nucleation and changes the aggregation process of hydrate crystal. Kinetic inhibitors mainly include two forms of surfactants and polymers [5, 6].

(1) Surfactants

Surfactants could reduce the surface tension of water, speed up the diffusion rate from gas to water and slow down the diffusion rate from gas to the hydrate crystal surface, so the hydrate growth would be controlled and inhibited. Surfactants mainly include polyoxyethylene nonylphenyl ester, sodium dodecylbenzenesulfonate, a mixture of 12–14 carboxylic acid and diethanolamine, polyglycerol oleate and the like.

(2) Polymers

The polymers are classified into amide polymers, ketone polymers, imine polymers and others. Amide polymers mainly include polyN-vinyl-epsilon-caprolactam, polyM-vinyl-epsilon-caprolactam, polyacrylamide, N-vinyl-N-methylacetamide and polymers containing diallylamine unit. It needs to note that most of the kinetic inhibitors have several deficiencies such as undesirable molecular structure, low inhibitory activity, poor universality, vulnerability to external influences and not suitable for the blocking caused by irregular well shut-in or insufficient inhibitor.

6.1.3 Anti-agglomerants

Anti-agglomerants, composed of polymers and surfactants, is mainly used as the oil-water emulsifier to disperse the water wrapped in oil phase into many small droplets. Hydrates are converted from these droplets and then solubilized in the emulsion solution, finally preventing the hydrates from coalescing and agglomerating and representing the lower hydrate blockage risks. Anti-agglomerants are most commonly used to work in case of the closed wellbore or large subcooling degree only when the oil and water coexist. Limited by their weak ability to disperse the water, the acting effect of anti-agglomerants is related to the content of water-in-oil and salt-in-water. That is, there is mutual selectivity between anti-agglomerants and oil-gas system, indicating the anti-agglomerants are probably not applicable to the hydrate prevention and management during the testing in a deepwater gas well.

6.2 Screening and Evaluation of Inhibitors

6.2.1 Alcohol Inhibitors Screening

The inhibitory effects of methanol, ethanediol and diethylene glycol on hydrate formation were studied by a set of high-pressure hydrates test equipment with the functions of mixing and agitating, which is constructed by China University of petroleum. The influence of alcohol types at a mass concentration of 5% and alcohol concentrations at 10 MPa on the phase equilibrium temperature of methane hydrate are shown in Table 6.2 and Fig. 6.1 respectively. At the mass concentration of 5%, methanol has the most effective inhibition, ethanediol less and diethylene glycol least. When the alcohol concentration is greater than 5%, every 1% growth of methanol concentration is accompanied by a 0.59 °C decrease in the hydrate phase equilibrium temperature and, each 1% increase of ethanediol and diethylene glycol

Table 6.2 Parameters and inhibitory effect of alcohol inhibitors

Experiment number	Inhibitors types	Mass concentration	Hydrate phase equilibrium point/(°C, MPa)	Hydrate Phase equilibrium temperature in pure water system/°C	Inhibitory effect/°C
C1	Methanol	5%	11.59, 10.05	13.39	1.8
C2	Ethanediol	5%	12.5, 10.07	13.4	0.9
C3	Diethylene glycol	5%	12.7, 10.03	13.37	0.67

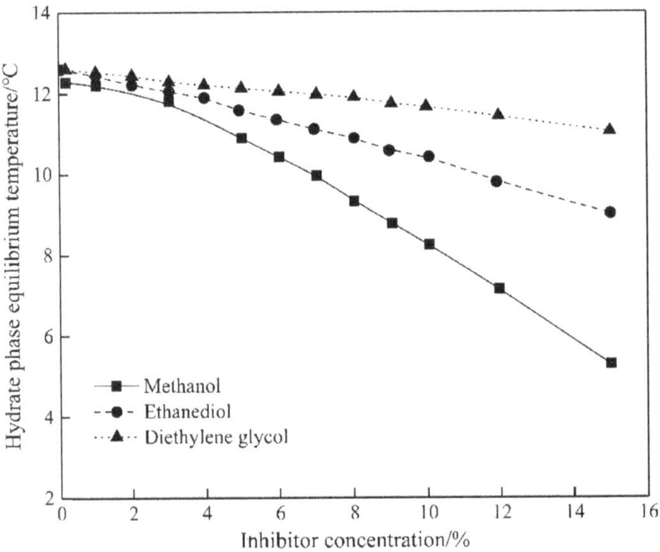

Fig. 6.1 Influence of alcohol inhibitors at different concentrations on the phase equilibrium temperature of methane hydrate

concentration reduces 0.23 and 0.1 °C of the hydrate phase equilibrium temperature, respectively. Besides, the relation between alcohol concentrations and the hydrate phase equilibrium temperature is approximately linear.

In summary, methanol has advantages such as low freezing point, strong water solubility, low costs, great inhibition effect. For the Ethylene glycol, it has advantages such as higher boiling point than methanol, low vapor pressure, high flash point, great inhibition effect and being environmentally friendly. Diethylene glycol has advantages such as high boiling point and flashing point, low vapor pressure, low volatility, small loss in the gas phase, but it has a worse inhibitory effect than methanol and ethanediol. Moreover, it is difficult to separate from the liquid hydrocarbon due to its large viscosity, causing Diethylene glycol is not used below −10 °C. Overall, methanol shall be preferred as the hydrate inhibition during well completion and well testing in deepwater.

6.2.2 *Salts Inhibitors Screening*

Common salt inhibitors mainly include NaCl, KCl, and $CaCl_2$, and their inhibitory effect is also experimentally evaluated using the high-pressure test equipment. The influence of inorganic salts types at a mass concentration of 5% and salts concentrations at 10 MPa on the phase equilibrium temperature of methane hydrate are shown in Tables 6.3 and 6.4 respectively. It illustrates that at a mass concentration of 5%, NaCl has the best inhibitory effect, while the inhibition effect of KCl and $CaCl_2$ is similar. Furthermore, the hydrate phase equilibrium temperature decreases with the salt concentration, so the system could withstand a higher subcooling degree before generating the hydrate. Concretely, the concentration of NaCl raise every 1%, the subcooling degree will increase about 0.55 °C, and the value will be about 0.45 °C for KCl and $CaCl_2$. According to this rule, the number of required salt inhibitors could be estimated on certain conditions.

Table 6.3 Parameters and inhibitory effect of inorganic salt

Experiment number	Inhibitors types	Mass concentration/%	Hydrate phase equilibrium point/(°C, MPa)	Phase equilibrium temperature of Hydrate in pure water system/°C	Inhibitory effect/°C
W1	NaCl	5	10.3,8.94	12.5	2.2
W2	KCl	5	10.4,8.21	11.7	1.3
W3	CaC12	5	10.4,8.21	11.7	1.3

Table 6.4 Influence of salt mass concentrations of salts on the hydrate phase equilibrium temperature

Salt Types	Hydrate phase equilibrium temperature at different salt concentrations/°C											
	1%	2%	3%	4%	5%	6%	7%	8%	9%	10%	12%	15%
NaCl	12.24	11.78	11.31	10.84	10.35	9.85	9.34	8.82	8.29	7.75	6.63	4.85
KCl	12.33	11.98	11.61	11.23	10.85	10.46	10.06	9.66	9.24	8.82	7.94	6.2
$CaCl_2$	12.33	11.97	11.6	11.22	10.84	10.45	10.05	9.64	9.22	8.78	7.9	6

6.2.3 Inhibitors Selection

In the system of thermodynamics inhibitors, both alcohols and salts could effectively reduce the hydrate phase equilibrium temperature. Unlike the thermodynamics inhibitors, the kinetic inhibitors mainly influence the hydrate formation rate, so the decrease of hydrate phase equilibrium temperature is not obvious. Currently, compound inhibitors have a better effect in the laboratory, whereas such inhibitors have not been fully verified in consideration of the environmental requirements.

Compared with salt inhibitors, alcohol inhibitors have lots of advantages such as small corrosion, full combustion with produced gas, and less environmental damage. Therefore, the methanol, having the best inhibition effect in alcohol inhibitors, is often used for hydrate prevention and management during on-site operation. However, it needs to note that in designing the pipe column and ground equipment, relevant sealing methods should not be ignored to satisfy the resistant methanol corrosion requirement for rubber parts.

6.3 Parameters Design During Inhibitors Injection

6.3.1 Injection System

For the methanol inhibitor, the typical injection system on the seabed is shown from Figs. 6.2, 6.3, 6.4, which mainly consists of a methanol tank, methanol injection pump, injection pipeline, and related interfaces. There are usually three injection positions on-site, and each of them is controlled by one injection pump.

Fig. 6.2 Methanol injection pump

Fig. 6.3 Methanol tank

6.3.2 *Injection Position*

During the deepwater gas well testing, there are three injection positions: the nozzle joint on the ground, near the mudline, and a definite depth below the mudline. The first two positions could effectively overcome the adverse effects caused by the low temperature at seabed and gas expansion. Besides, to ensure that injected inhibitors can cover all the well sections in which hydrate is possiblely generated, the deepest position located in the hydrate formation area is always taken as the injection point.

(1) One definite depth below the mudline

When the water depth is more than 1500 m, the temperature near the mudline is usually 2–4 °C. It is assumed that the temperature gradient below the mudline is 3 °C/100 m, and the critical depth without hydrate formation is 600 m below mudline according to the critical hydrate phase equilibrium temperature of 21 °C. So, inhibitor injection shall be set at this position, and the injection amount should account for half of the total injection amount.

(2) Near the mudline

To prevent the hydrate formation in the low-temperature environment near the mudline, one injection point should be established near the mudline and the injection amount should account for 25% of the total injection amount.

(3) Nozzle joint on the ground

Under the throttle effect of the nozzle joint, the downstream pressure would be decreased, causing the gas expansion and the fluid temperature to reduce. To prevent hydrates formation in the ground pipeline, one injection point should be also established at a wellhead and nozzle joint. The injection amount is equally 25% of the total injection amount.

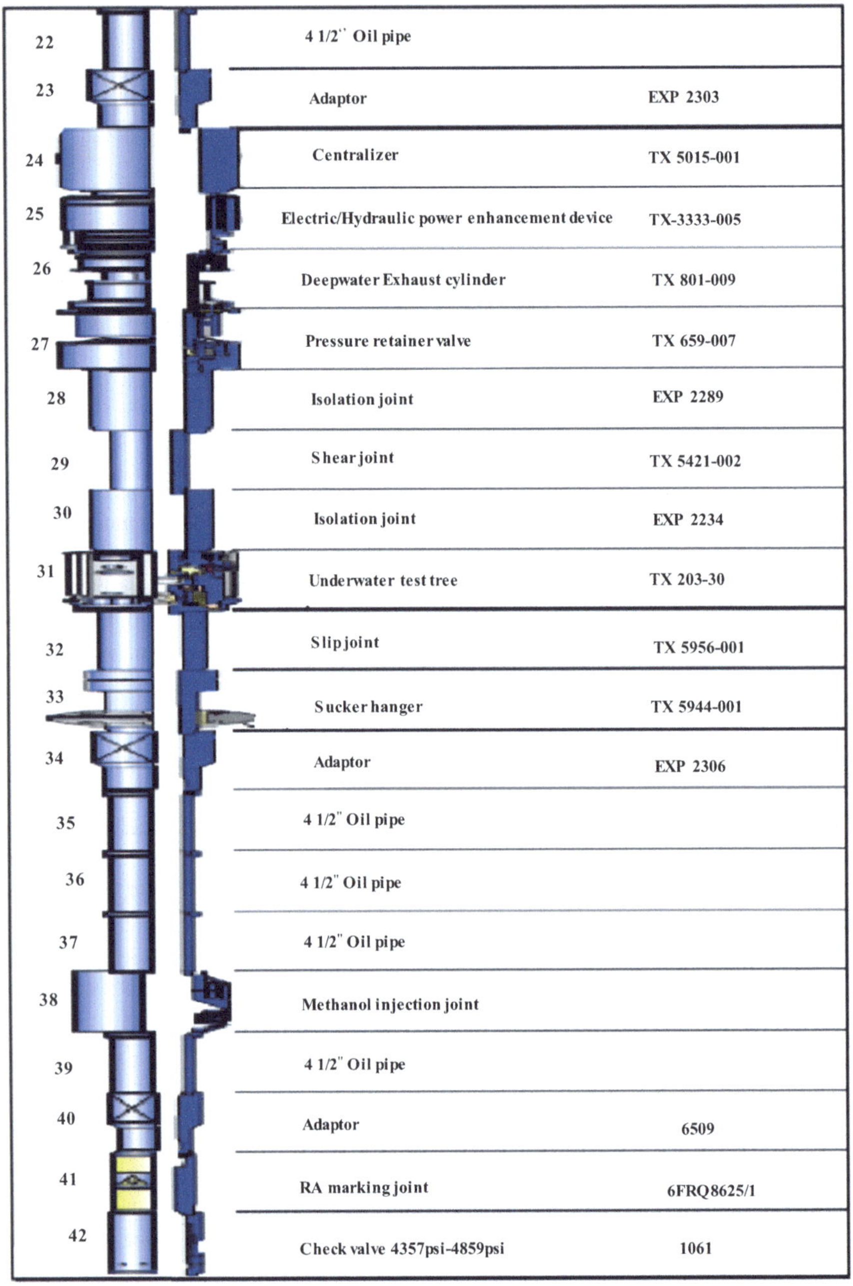

Fig. 6.4 Injection pipeline and related interfaces

6.3.3 Injection Pressure

Inhibitors are injected into the wellbore by the related injection line and injection pump on the ground. Determining the injection pressure could provide a certain reference for the selection of the injection pump. In practical operation, the injection pressure in different test conditions needs to be calculated according to wellbore pressure at the injection point. For methanol and ethanediol, the minimum injection pressure is shown in Eq. 6.1.

$$P_s \geq P_z + \Delta P_f + \Delta P_j - \Delta P_h \tag{6.1}$$

$$\Delta P_f = \frac{\lambda \rho v^2}{2D} \cdot H \tag{6.2}$$

$$\Delta P_h = \rho g H \tag{6.3}$$

where P_s shows the minimum injection pressure of injection pump, MPa; P_z shows the wellbore pressure at injection position, MPa; ΔP_f refers to the frictional pressure drop of injection pipeline, MPa, calculated by Eq. 6.2; ΔP_j refers to the partial pressure loss at injection position, MPa; ΔP_h is the hydrostatic pressure of inhibitors, MPa, calculated by Eq. 6.3; λ stands for the friction coefficient; ρ stands for the inhibitor density, kg/m^3; v represents the injection velocity of inhibitor, m/s; D is the diameter of injection pipeline, m; H shows the injection position, m.

6.3.4 Injection Concentration

During the deepwater gas well test, the appropriate injection concentration shall be determined based on the inhibitory effect and economic performance of methanol. The flow chart for determining the injection concentration of methanol is shown in Fig. 6.5. Notice that there is a difference between the calculation process and actual operation, so it is necessary to multiply a safety factor to render the calculation closer to reality.

6.3.5 Injection Rate

The injection rate of inhibitor is related to factors such as condensate water, mud filtrate, water production rate, gas expansion during its upward flow and inhibitor loss in the gas phase, which is calculated by Eq. 6.4. Due to the position uncertainty for water production when the well testing starts, the high injection rate should be generally established to avoid the hydrate formation.

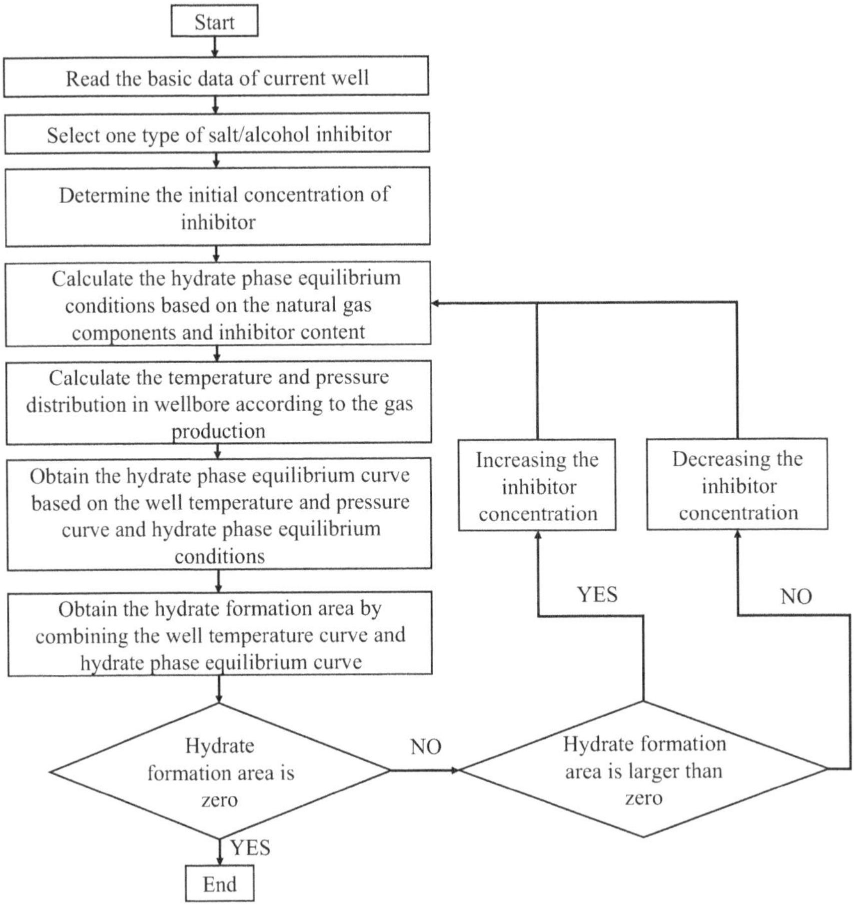

Fig. 6.5 Flow chart for determining the injection concentration during deepwater gas well testing

$$q_s \geq \frac{c}{1-c} q_w + q_l \tag{6.4}$$

$$q_l = 1.97 \times 10^{-2} P^{-0.7} \exp\left(6.054 \times 10^{-2} T - 11.28\right) \tag{6.5}$$

where q_s shows the injection rate of inhibitor, L/min; q_w is the water production rate, L/min; q_l is the loss rate of inhibitor in the gas phase, L/min, which can be calculated by Eq. 6.5; c stands for the injection concentration of inhibitor, %; P is the system pressure, MPa; T is the system temperature, K.

6.3.6 *Injection Amount*

6.3.6.1 Depression of Hydrate Formation Temperature

The hydrate formation temperature would reduce after injecting the chemical inhibitor. Hammerschmidt [7] first proposed the semi-empirical relationship between the depression of hydrate formation temperature (ΔT) and percentage concentration of inhibitor solution ($W\%$) shown in Eq. 6.6. Based on the Equation, the relation between methanol content and ΔT is analyzed and shown in Table 6.5. It can be seen that the prediction deviation by Eq. 6.6 is larger when the mass fraction of methanol is greater than 30%. Nielsen et al. also pointed out that the semi-empirical equation was only applicable to the inhibitor solution with a mass fraction less than 20%. In high methanol content, the Nielsen-Bucklin modified formula is often used to calculate the ΔT, so the methanol consumption is estimated by Eq. 6.7. Based on the equation, the methanol consumption of one gas field in northern Shanxi is predicted as shown in Table 6.6, indicating that the deviation between the predicted value and experimental value is small.

Table 6.5 Relation between methanol content and ΔT

Percentage concentration of methanol $W/\%$	$\Delta T/°C$		
	Experimental result	Calculation result by Hammerschmidt's semi-empirical Equation	Deviation
10.0	3.1	4.3	1.2
20.0	10.0	9.6	0.4
30.0	12.2	16.4	4.2
35.0	15.8	16.8	1.0
50.0	23.1	38.4	15.3
65.0	40.4	71.3	30.9

Table 6.6 Prediction for methanol consumption of one gas field in Northern Shanxi

Operation condition		Hydrate formation temperature without inhibitor/°C	$\Delta T/°C$	Percentage concentration of methanol $W/\%$		
P/MPa	T/°C			Calculation value	Experimental value	Deviation
4	0.5	11.0	10.5	25.6	25.7	0.3
8	0.5	15.5	15.0	35.5	35.1	−1.1
4	−9.0	11.0	20.0	45.8	45.7	−0.2
8	−9.0	16.4	25.4	56.0	55.3	−1.3
17	−9.0	22.0	31.0	66.1	66.4	0.5
22	−9.0	24.5	33.5	70.3	70.1	−0.3

$$\Delta T = \frac{KW}{M(100 - W)} \tag{6.6}$$

$$W = 1.89\left(1 - e^{-0.013886\Delta T}\right) \tag{6.7}$$

where M is the relative molecular mass of inhibitor; W is the mass fraction of inhibitor solution, %; ΔT represents the depression of hydrate formation temperature, °C; K shows one constant related to inhibitor; taking 1228 for methanol, ethanediol, isopropanol, ammonia; taking 1200 for calcium chloride and taking 2425 for diethylene glycol.

6.3.6.2 Determination of Injection Amount

According to the well testing plan on-site, the required injection amount of inhibitor should be determined in advance to make the necessary preparations. The inhibitor consumptions mainly include the amount to shift the hydrate phase equilibrium curve and the amount that saturated gas requires. The saturated vapor pressure of the electrolyte solution is usually lower than that of condensed water, so the amount of inhibitor in the vapor state is often ignored. Herein, the unit consumptions of the inhibitor could be calculated by Eq. 6.8. When using alcohol inhibitors, the consumption converting to gas phase must be taken into account. If the concentration of the recovery solution is given, the unit consumptions of methanol could be determined by Eq. 6.9.

$$q_s = \frac{(W_1 - W_2)K}{C - K} \tag{6.8}$$

$$q_c = \frac{W_1 - W_2}{C - K} + K \cdot 10^{-3} \cdot \alpha \tag{6.9}$$

$$\alpha = 1.97 \times 10^{-2} P^{-0.7} \exp\left(6.054 \times 10^{-2} T - 11.128\right) \tag{6.10}$$

where W_1 is the water content of natural gas at the injection point of inhibitor, g/cm^3; W_2 is the final water content of fluid at the endpoint, g/cm^3; C, K refer to the weight concentration of added and recycled inhibitor, %; α represents the amount of methanol converting to the gas phase, calculated by Eq. 6.10; P is the pressure, MPa; T is the temperature, K.

6.3.6.3 Design Chart for Injection Amount

The inhibitory effect of inhibitor is mainly related to its type and concentration. The temperature and pressure for hydrate formation after injecting the methanol and ethanediol with different concentrations is shown in Table 6.7, which could

Table 6.7 Temperature and pressure for hydrate formation at different methanol and ethanediol concentrations

Pressure/MPa	Hydrate formation temperature under different methanol concentrations/°C					Hydrate formation temperature under different ethanediol concentrations/°C					
	10%	20%	30%	40%	50%	10%	20%	30%	40%	50%	60%
20.684	10.9	5.0	3.1	−8.9	−17.5	14.0	10.8	6.5	0.9	−6.3	−15.3
18.184	10.4	4.5	2.6	−9.4	−17.9	13.5	10.2	6.0	0.5	−6.6	−15.6
15.684	9.9	4.0	2.1	−9.8	−18.4	12.9	9.7	5.4	0.0	−7.0	−16.0
13.184	9.2	3.4	1.5	−10.2	−18.8	12.2	9.1	4.9	−0.5	−7.5	−16.4
11.684	8.7	3.0	1.1	−10.5	−19.0	11.8	8.6	4.5	−0.8	−7.9	−16.6
10.184	8.2	2.5	0.6	−10.9	−19.3	11.2	8.1	4.0	−1.3	−8.1	−16.8
8.684	7.5	2.0	0.1	−11.3	−19.5	10.5	7.4	3.3	−1.9	−8.5	−17.0
7.184	6.6	1.0	−0.8	−11.9	−19.9	9.5	6.5	2.5	−2.7	−9.3	−17.5
6.434	6.0	0.5	−1.3	−12.3	−20.2	8.9	5.9	1.9	−3.2	−9.6	−17.9
5.684	5.3	−0.2	−2.0	−12.9	−20.8	8.1	5.1	1.2	−3.8	−10.3	−18.4
4.934	4.4	−1.1	−2.8	−13.7	−21.5	7.3	4.3	0.4	−4.6	−10.9	−19.0
3.434	2.0	−3.4	−5.2	−15.9	−23.5	4.8	1.9	−2.0	−6.9	−13.1	−21.0
2.684	0.1	−5.3	7.0	17.6	25.3						

provide a theoretical reference for determining the injection amount of inhibitor. To consult the methanol injection amount under different conditions quickly and conveniently, a design chart for "Methanol Injection Amount and Hydrate Formation Temperature-pressure" is made and shown in Fig. 6.6.

6.4 Hydrate Blockage Management Method Based on Incomplete Inhibition

At present, injecting excessive inhibitors is usually performed to completely prevent hydrate formation during deepwater drilling, which has many weaknesses such as large consumption, high costs, and non-environmental friendliness. Moreover, the required injection rate may exceed the equipment capability under large water yield. Now, to avoid the high consumption of inhibitors and reduce the equipment requirements for injection and storage, Wang et al. and Sun et al. [8–13] proposed one hydrate blockage management method based on the HBFW and well testing system.

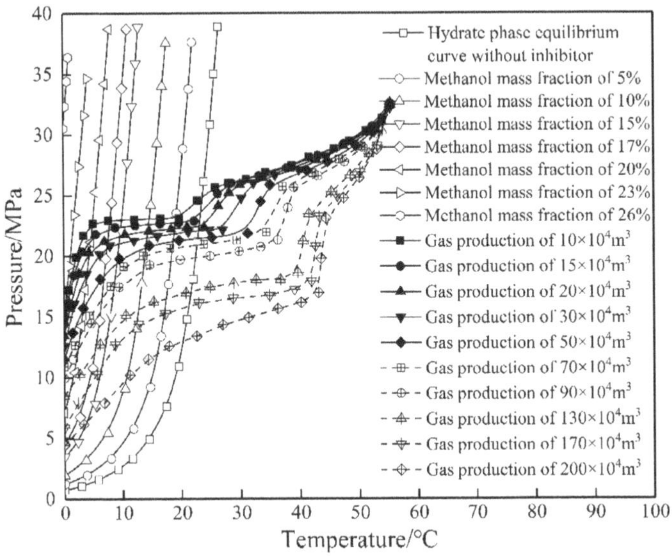

Fig. 6.6 Design chart for "methanol injection amount and hydrate formation temperature-pressure"

6.4.1 Hydrate Blockage Management Based on the HBFW

6.4.1.1 Management Procedures

Hydrate blockage management method based on HBFW allows a small amount of hydrates to be generated, and the limited consumption of inhibitors is determined in terms of the evolution characteristics of hydrate migration, deposition, and blockage in the wellbore. Compared with the complete inhibition method, this method could significantly reduce the inhibitors consumption and its specific management procedures are as follows [12–15]:

(1) Calculate the well temperature and pressure fields;
(2) Determine the hydrate formation area based on the theory on hydrate thermodynamic equilibrium;
(3) Calculate the hydrate formation rate based on the theory on hydrate formation kinetics;
(4) Calculate the hydrate deposition rate and δ_D distribution and determine the spatial and temporal variation of hydrate blockage in well, together with HBFW;
(5) By comparing HBFW with operation time, determine whether to proceed with the inhibitor injection or not, then determine the injection concentration and stipulate related measures to manage the hydrate blockage.

Table 6.8 Basic parameters of case

Basic parameters	Value	Basic parameters	Value
Well depth/m	3500	Water depth/m	1500
Heat conductivity coefficient of Stratum/W $(m \cdot k)^{-1}$	2.2	Heat conductivity coefficient of steel/W $(m \cdot K)^{-1}$	43.2
Specific heat of stratum rock/J $(kg \cdot K)^{-1}$	830	Temperature on Sea bed/°C	2.9
Stratum rock density/$kg \cdot m^{-3}$	2640	Stratum temperature at test position/°C	52.9
Heat conductivity coefficient of seawater/W $(m \cdot k)^{-1}$	1.73	Relative density of natural gas	0.554
Seawater specific heat/J $(kg \cdot K)^{-1}$	3890	Testing wellbore/mm	114.3
Seawater density/$kg \cdot m^{-3}$	1025	Heat conductivity coefficient of well annulus fluid/W $(m \cdot K)^{-1}$	0.6
Heat conductivity coefficient of cement ring/W $(m \cdot k)^{-1}$	0.35	Water production/$m^3 \cdot d^{-1}$	15
Gas flow/10^4 $m^3 \cdot d^{-1}$	30	Downhole pressure/MPa	26

6.4.1.2 Case Application

For the hydrate blockage management method based on HBFW, its specific application is further illustrated by one case and the basic parameters of the case well are shown in Table 6.8.

(1) Calculate the well temperature and pressure fields

The hydrate phase equilibrium curves at different methanol concentrations are shown in Fig. 6.7. It can be seen that the hydrate phase equilibrium curve would increasingly shift towards the upper left orientation with the inhibitor concentration, and the shifting increment is greater continuously. The well temperature distribution calculated by the well temperature field model is shown in Fig. 6.8 [12, 16]. The seawater temperature gradually decreases with well depth and then reaches a minimum value of 2.9 °C at mudline, while the stratum temperature at well-testing points is up to 52.9 °C. With the upward migration of well fluid, the fluid temperature would gradually reduce and it would drop to the minimum value of 12.8 °C when the fluid reaches the 170 m below sea surface due to the continuous heat transfers between fluid and surrounding stratum or seawater.

(2) Determine the Hydrate Formation Area in Wellbore

The hydrate formation area at different methanol concentrations is shown in Fig. 6.8 and Table 6.9 [12, 16], illustrating that the hydrate formation area is 0–1000 m when no methanol is injected. The hydrate formation area would become smaller after inhibitor injection and the methanol concentration over 30% makes the hydrate formation area disappear. As a general rule, for the hydrate blockage management, injecting the inhibitor with high concentration (more than 30%) is usually performed

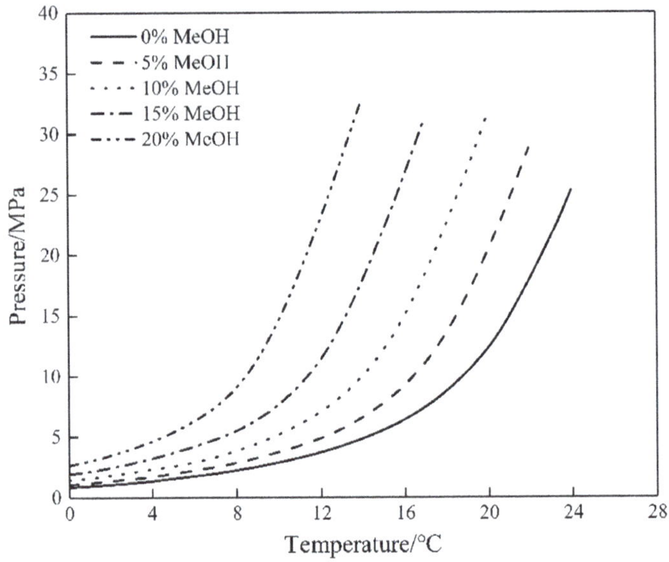

Fig. 6.7 Hydrate phase equilibrium curve at different methanol concentrations

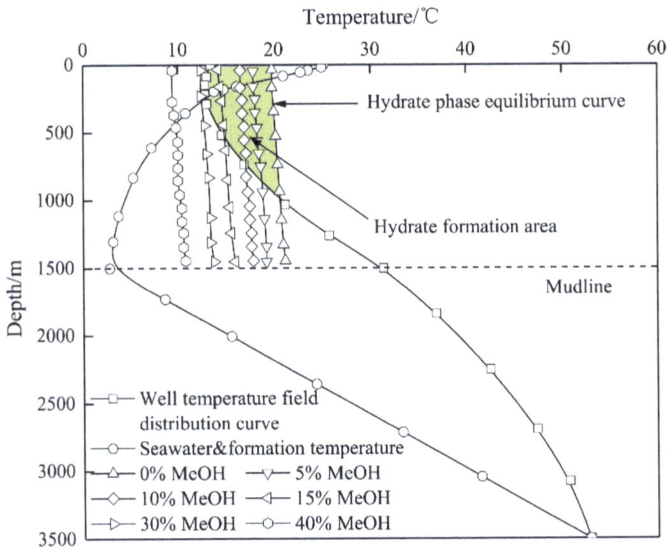

Fig. 6.8 Well temperature distribution and hydrate formation area

Table 6.9 Hydrate formation area at different methanol concentrations

Methanol concentration/%	Hydrate formation area/m	Position of maximum subcooling degree/m
0	0–1000	150
5	0–890	150
10	0–770	150
15	0–570	150
>30	N/A	150

to make the well temperature and pressure field fully outside the hydrate formation area, so there is no need to consider the various characteristics of hydrate migration, deposition and blockage in well. However, unlike this traditional method, these characteristics must be taken into account using the hydrate management method based on HBFW. Only in this way can we successfully design and popularize the corresponding hydrate management scheme.

(3) Determine the δ_D Distribution

The δ_D distribution with well testing time at different methanol concentrations is shown in Fig. 6.9 [12, 16]. It can be seen that δ_D increases with operation time, eventually leading to well hydrate blockage. Furthermore, the growth rate of the hydrate layer would significantly reduce and HBFW would become larger with the inhibitor concentration.

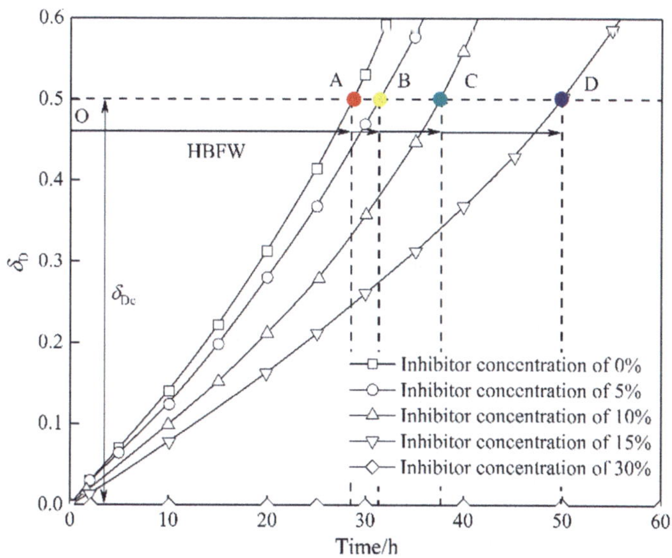

Fig. 6.9 δ_D Distribution with well testing time after inhibitor injection

(4) Determine the Spatial and Temporal Variation of Hydrate Blockage in Wellbore, Together with HBFW

As shown in Fig. 6.9, without inhibitor injection, δ_D increases fastest and the hydrate blockage occurs when the testing operation time reaches about 28.6 h. That is to say, the HBFW is 28.6 h. After inhibitor injection, the growth rate of δ_D significantly slows down, changing the spatial and temporal characteristics of hydrate blockage. Accordingly, the HBFW increases from 28.6 h to 31.6 h, 37.5 h and even infinite with inhibitor concentration.

(5) Determine whether to inject inhibitor or not, then determine the injection concentration

If well testing time is less than the HBFW, the testing operation could still be carried out smoothly. Otherwise, inhibitors need to be injected timely to delay the hydrate blockage. For example, suppose the planned testing time is 40 h. If methanol concentration of 15% is selected, its injection will render the HBFW expand to 50.1 h and be able to ensure the operation safety, so that the injection of methanol with excessive concentration will not be needed. Obviously, the hydrate blockage management method based on incomplete inhibition could greatly reduce and save the amount of inhibitor, further lowering the standards of equipment injection and platform reserve.

6.4.2 Hydrate Blockage Management Based on the Well Test System

To complete the downhole sampling and obtain the well testing dates, it is necessary to change the gas production during deepwater gas well testing. What's more, the hydrate blockage could be prevented and managed by altering the well testing sequence. To be more specific, under the low testing production, the heat exchange between testing fluid and external environment is more sufficient, so it is easy for the well temperature and pressure to enter the hydrate formation area, greatly facilitating the generation of hydrate layer on pipewall. Moreover, when the testing production exceeds one critical value, the well temperature and pressure can not meet the condition of hydrate phase equilibrium and the hydrate layer on pipewall will begin to decompose, finally achieving the purpose of hydrate blockage management without inhibitor injection [8].

In normal well test sequence, the test operation is generally carried out in a positive sequence from low production to high production (taking 30, 50, 80, 110 $\times$ 10^4 m^3/d as an example), which is as shown in Fig. 6.10a. In the testing production of 30, 50 $\times$ 10^4 m^3/d, both hydrate formation area and hydrate blockage risks exist in well. However, after the testing production of 80 $\times$ 10^4 m^3/d is adjusted to the second test sequence and the testing production of 50 $\times$ 10^4 m^3/d is adjusted to the third test sequence, the well temperature and pressure would be outside the hydrate formation

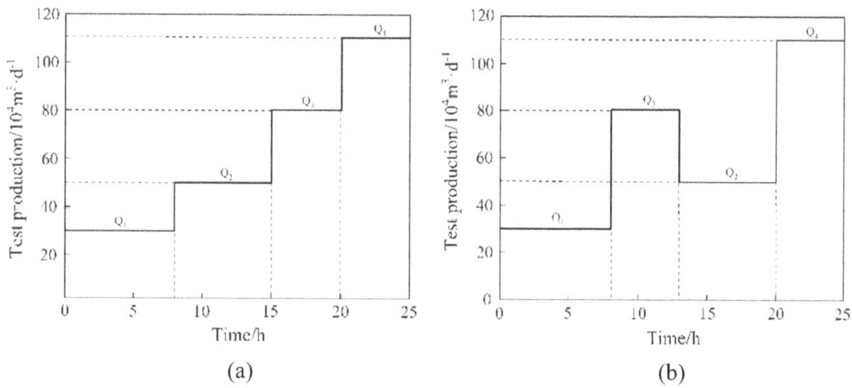

Fig. 6.10 Well test system, **a** normal well test sequence, **b** adjusting the well test sequence

area. Consequently, the hydrate layer generated in previous testing production would begin to decompose, which is shown in Fig. 6.10b.

The well hydrate blockage risks in test system of normal test sequence and adjusting the test sequence are shown in Fig. 6.11 [8], illustrating the value of δ_D increases firstly and then decreases with test time in the test system of the normal test sequence. Moreover, the maximum shrink diameter ratio of wellbore reaches 57%, indicating the hydrate blockage is much more likely to happen in wellbore. However, the maximum shrink diameter ratio of wellbore has fallen from 57% to 26% by adjusting the well test sequence. This is mainly because the temperature and pressure conditions for hydrate formation no longer exist in wellbore under 80×10^4

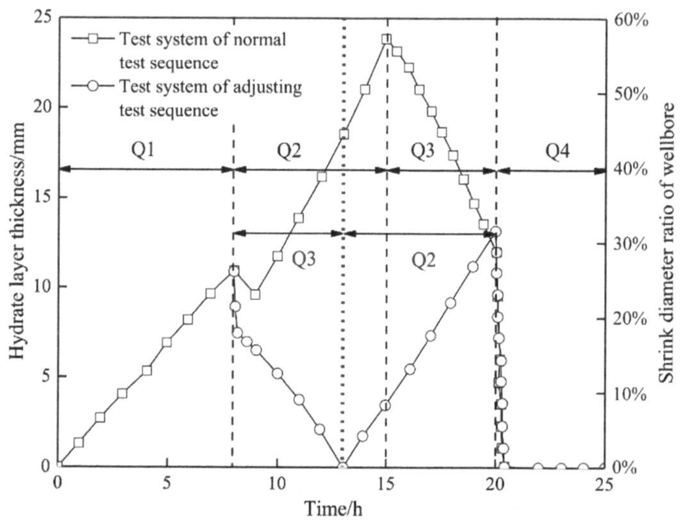

Fig. 6.11 Hydrate blockage risks in different test systems

m^3/d, leading to the gradual decomposition of hydrate layer generated under 30 × 10^4 m^3/d. Afterwards, the hydrate layer would grow again when the gas production falls below the critical gas production (50 × 10^4 m^3/d). Similarly, the hydrate layer converted in the third test sequence (50 × 10^4 m^3/d) would begin to decomposition in the fourth test sequence (110 × 10^4 m^3/d), ensuring that the whole well testing is carried out successfully. Noticeably, the hydrate layer decomposition rate under 110 × 10^4 m^3/d is far larger than that under 80 × 10^4 m^3/d, because the higher gas production could result in higher well temperature and greater driving force for hydrate decomposition. Also, the fluid shearing force applied to the hydrate layer would increase with gas production, possibly removing part of the hydrate layer [17, 18]. Given the lack of research in this area currently, the fluid shearing force is always not considered during the simulation.

Reasonably adjusting the well test sequence could reduce significantly the hydrate blockage risks in the wellbore, which has advantages such as low costs, environmental friendliness, simplicity, and efficiency. However, it also maybe change the methods about test analysis and reservoir evaluation, so how to implement it smoothly needs further research.

6.5 Risk Early-Warning of Hydrate Blockage

6.5.1 Monitoring Device

The temperature and pressure in wellbore have unique characteristics due to the hydrate deposition and blockage. By monitoring these data, the spatial and temporal characteristics of hydrate blockage and its risk severity could be calculated during the deepwater gas well test based on the hydrate blockage model, so the corresponding risk-warning could be carried out in advance. After the injection position and injection rate of the inhibitor are determined, the security and efficiency of the test are finally guaranteed.

Hydrate layer on pipewall keeps thickening with hydrate particles deposition, and the frictional pressure drop consists of a relatively stable stage, slow rise, severe fluctuation stage, and rapid increase. The former two stages are the early period of hydrate blockage, in which the hydrate formation and deposition have little influence on the well testing, so that the test operation could be still carried out. Herein, the severity of hydrate blockage could be determined by risk early-warning. In addition, the latter two stages are the later period of hydrate blockage, in which the well testing cannot be performed normally. Herein, the hydrates deposition alternates with hydrates removal, causing the drastic fluctuation of frictional pressure drop.

A set of device for the early monitoring of the hydrate blockage during the deepwater gas well test, shown in Fig. 6.12 [19, 20], is proposed based on the pressure drop characteristics in the wellbore. The device is mainly composed of data acquisition and transmission system, platform control system and automatic injection of

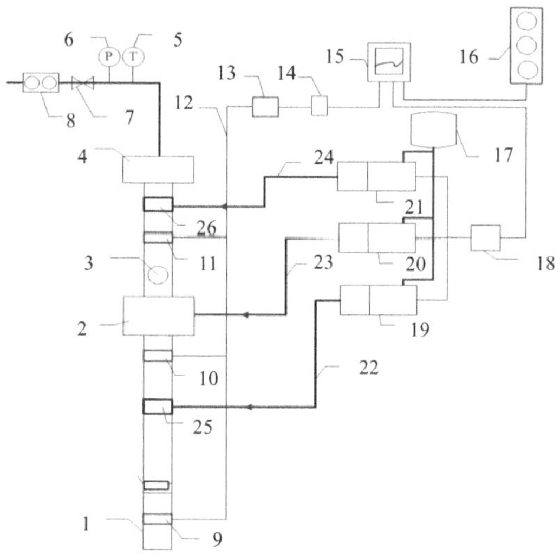

Fig. 6.12 Device for early monitoring the hydrate blockage, 1—deepwater gas well, 2—underwater testing tree, 3—blowout preventer valve, 4—flow control head, 5—wellhead thermometer, 6—wellhead pressure gauge, 7—injection nozzle, 8—wellhead flowmeter, 9—the first sensor group of temperature and pressure, 10—the second sensor group of temperature and pressure, 11—the third sensor group of temperature and pressure, 12—optical cable, 13—optical interface, 14—photoelectric demodulator, 15—computer, 16—alarm device, 17—inhibitor tank, 18—signal actuator; 19—the first pump for inhibitor injection, 20—the second pump for inhibitor injection, 21—the third pump for inhibitor injection, 22—the first injection pipeline, 23—the second injection pipeline, 24—the third injection pipeline, 25—the first injection joint, 26—the second injection join

inhibitor system. The data acquisition and transmission system could detect the real-time well temperature-pressure data and transmit it to the platform control system. The platform control system is mainly used to estimate the severity and dimensions of hydrate blockage, when and where the hydrate blockage occurs, and send out corresponding early-warning signals. According to these signals, the injection position, injection rate and injection concentration of inhibitor are determined. Finally, based on the injection requirements, the automatic injection of the inhibitor system starts to automagically inject inhibitor into the well by injection pump.

6.5.2 *Early-Warning Flow*

After monitoring the data such as flow volume, temperature and pressure in bottom hole and wellhead, basic parameters of the wellbore, seawater and stratum temperature, we could further simulate the time and space distribution of hydrates in well and evaluate the severity and dimensions of hydrate blockage risk. The early-warning flow is as below.

(1) Measure the gas production (Q_g) and water production (Q_w) by flowmeter in the wellhead, and measure the fluid temperature (T_{wh}) and pressure (P_{wh}) in wellhead by thermometer and pressure gauge respectively;
(2) Calculate the well temperature and pressure distribution according to the multi-phase flow model and determine the hydrate formation area in wellbore based on the hydrate phase equilibrium theory;
(3) Evaluate the spatial and temporal characteristics of hydrate blockage and its risk severity;
(4) Determine the early-warning level in terms of the risk severity in the wellbore.

According to the shrink diameter ratio of the wellbore, the early-warning signal could be divided into four levels. The warning level 1 is carried out with a well shrink diameter ratio of 70%–90%. Again, the warning level 2 and 3 are executed with a shrink diameter ratio of 60%–70% and 40%–60%, finally the warning level 4 is carried out when the shrink diameter ratio is less than 40%.

6.5.3 Field Application

According to the early-warning flow above, the hydrate blockage risk during testing, in well #3 located in one certain sea area, is evaluated and early warned. The change of well pressure drop during the development of hydrate blockage is shown in Fig. 6.13, illustrating the process of hydrate blockage could be divided into four stages: small amount of hydrate formation, hydrate formation, and deposition, repeated deposition and removal of hydrates on pipewall and hydrate blockage in the wellbore. The δ_D, closely related to the pressure drop, could represent the severity and dimensions of hydrate blockage. So, based on the shrink diameter ratio of well, the hydrate blockage

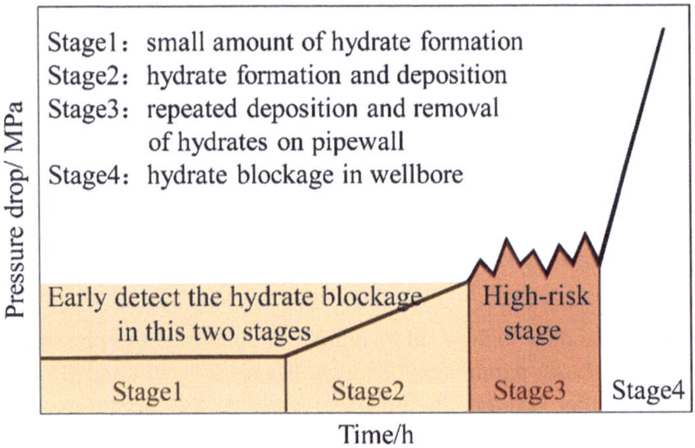

Fig. 6.13 Change of well pressure drop during the development of hydrate blockage

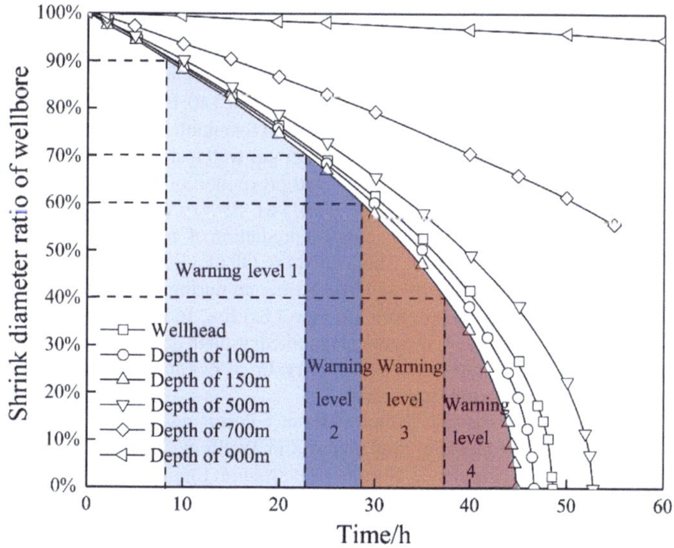

Fig. 6.14 Early monitoring results for hydrate blockage

risks could be timely detected and monitored at the first and second stages and the corresponding early-warning signal could be sent out, which is shown in Fig. 6.14.

All in all, through the risk early-warning, we could monitor the hydrate blockage risk timely and find out when and where it will happen, finally send out the early-warning signal. Given this, actions related to plug removal should be taken immediately to avoid the interruption of well testing. Therefore, for hydrate management, it is instructive to investigate the risk early-warning of hydrate blockage.

References

1. Li G, Li X, Tang L et al (2007) Experimental investigation of production behavior of methane hydrate under ethylene glycol injection in unconsolidated sediment. Energy Fuels 21(6):3388–3393
2. Chen Z, Li Q, Yan Z et al (2010) Phase equilibrium and dissociation enthalpies for cyclopentane + methane hydrates in nacl aqueous solutions†. J Chem Eng Data 55(10):4444–4449
3. Chen Z, Feng J, Li X et al (2014) Preparation of warm brine in situ seafloor based on the hydrate process for marine gas hydrate thermal stimulation. Ind Eng Chem Res 53(36):14142–14157
4. Du J, Wang X, Liu H et al (2019) Experiments and prediction of phase equilibrium conditions for methane hydrate formation in the NaCl, $CaCl_2$, $MgCl_2$ electrolyte solutions. Fluid Phase Equilib 479:1–8
5. Karisiddaiah SM (2003) Report: fourth international conference on gas hydrates. Held at Yokohama, Japan, 19–23 May 2002
6. Chen G, Sun C, Qinglan MA (2008) Gas hydrate science and technology. Chemical Industry Press

7. Hammerschmidt EG (1934) Formation of gas hydrates in natural gas transmission lines. Ind Eng Chem 26(8):851–855
8. Zhang J, Wang Z, Sun B et al (2019) An integrated prediction model of hydrate blockage formation in deep-water gas wells. Int J Heat Mass Transf 140:187–202
9. Wang Z, Zhao Y, Sun B et al (2016) Heat transfer model for annular-mist flow and its application in hydrate formation risk analysis during deepwater gas well testing. J Hydrodyn 31(1):20–27
10. Wang Z, Zhao Y, Sun B et al (2018) Features and prevention of gas hydrate blockage in test strings of deep-water gas wells. Nat Gas Ind 38(01):81–88
11. Wang Z, Yu W, Meng W et al (2018) Quantitative prediction of gas hydrate blockage in testing tubing for deepwater gas well testing. China Offshore Oil Gas 30(03):122–131
12. Wang Z, Yang Z, Zhang J et al (2018) Flow assurance during deepwater gas well testing: hydrate blockage prediction and prevention. J Petrol Sci Eng 163:211–216
13. Zhao Y, Wang Z, Yu J, et al (2017) Hydrate plug remediation in deepwater well testing: a quick method to assess the plugging position and severity. In: SPE Annual technical conference and exhibition. Society of Petroleum Engineers
14. Deng Z, Wang Z, Zhao Y et al (2017) Flow assurance during gas hydrate production: hydrate regeneration behavior and blockage risk analysis in Wellbore. In: Abu Dhabi international petroleum exhibition and conference. Society of Petroleum Engineers
15. Zhao Y, Wang Z, Zhang J, et al (2016) Flow assurance during deepwater gas well testing: when and where hydrate blockage would occur. In: SPE annual technical conference and exhibition. Society of Petroleum Engineers
16. Wang Z, Yang Z, Zhang J, et al (2018) Quantitatively assessing hydrate-blockage development during deepwater-gas-well testing. SPE J 23(4):1166–1183
17. Lingelem MN, Majeed AI, Stange E (2010) Industrial experience in evaluation of hydrate formation, inhibition, and dissociation in pipeline design and operation. Ann N Y Acad Sci 715(1):75–93
18. Sloan JED, Koh CA (2007) Clathrate hydrates of natural gases. CRC press
19. Wang Z, Zhao Y, Sun B et al (2016) Early monitoring device and method for hydrate blockage in production pipeline of deepwater gas well: 201610735711.2
20. Wang Z, Zhao Y, Sun B et al (2016) Monitoring device and method for gas hydrate blockage during deepwater gas well testing: 201610735694.2

Chapter 7
Hydrate Management Software in Deepwater Gas Well and Case Analysis

The "Hydrate Management Software in Deepwater Gas Well" is developed by combining the hydrate formation area prediction model, the hydrate deposition and blockage models and hydrate management methods established above. For the production and test well in one certain sea area, the developed software could be used to predict their hydrate formation area and hydrate blockage risk, eventually providing the support for determining the injection parameters of inhibitor.

7.1 Software Overview

The "Hydrate Management Software in Deepwater Gas Well" is used to predict the hydrate formation area under various working conditions and optimize the injection parameters of single inhibitor or combined inhibitors, which has important significance for guiding the hydrate prevention and management on-site [1–4].

In this software, the related data is saved in the form of a database, which has the unique characteristics such as simple operation, convenient input, and unified storage or separate storage of calculation results. The software is mainly composed of five modules shown in Fig. 7.1: data input module; calculation module of hydrate phase equilibrium conditions in different gas components, inhibitor types and inhibitor concentrations; calculation module of well temperature and pressure fields under various work conditions; prediction module of hydrate formation area in wellbore under various work conditions; and optimization module of injection parameters.

The software, programmed by VB.net and having an access database, totally includes six main menus: file, data input, function, tool, help, and exit. Its output forms include text and graphics. The software could operate under in 10/Windows9x/NT/2000/XP/Me/7 system and offer the independent operation functions of delete, add, save, modify for database, which has the advantages such as beautiful interface, reasonable design and easy to use.

© Science Press and Springer Nature Singapore Pte Ltd. 2020

Z. Wang et al., *Natural Gas Hydrate Management in Deepwater Gas Well*,
https://doi.org/10.1007/978-981-15-6418-5_7

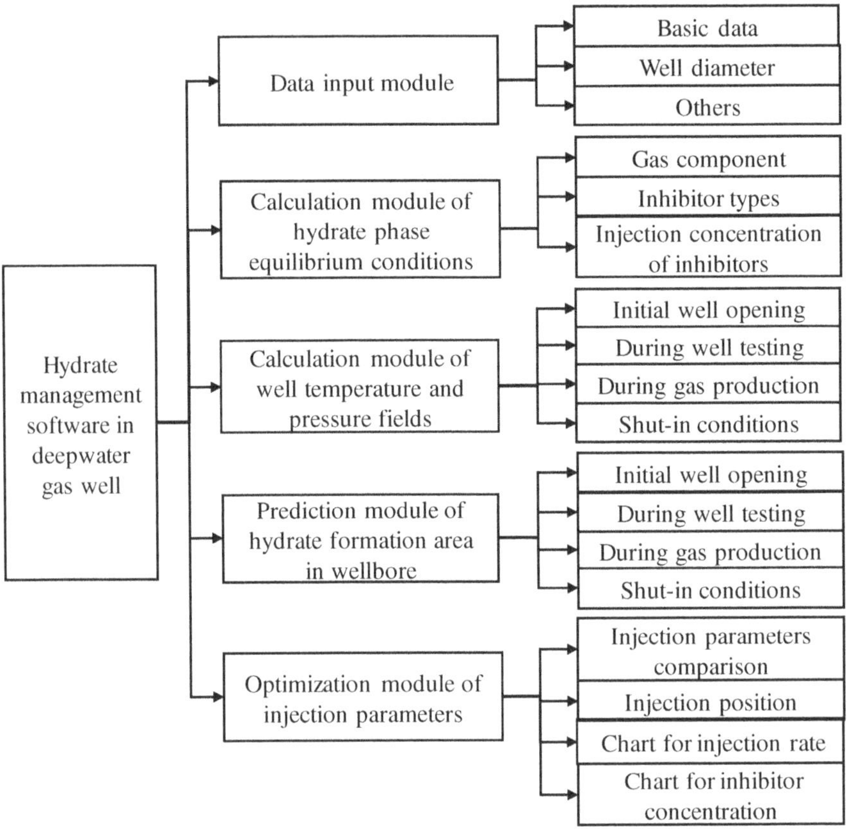

Fig. 7.1 Composition of hydrate management software in deepwater gas well

7.2 Software Composition

7.2.1 Data Input Module

The data input module is mainly used for the input, modification and save of data every well, which provides necessary data for other calculation modules. This module includes five parts: basic data, well structure, well deviation, pipe column data, and temperature parameters. The input parameters of each part are listed in Table 7.1.

7.2.1.1 Basic Data

After opening or building a new well, the software would automatically access to the default option-"Basic Data" shown in Fig. 7.2, where the user could input or

Table 7.1 data input and related parameters

Module composition	Input parameters
Basic data	Well name, well category, well position coordinates, geographic position, structure position, well type, drilled layers and main targeted layer; well depth, datum plane of well depth, kelly bushing, seawater depth, inner diameter of water riser, seawater density, reservoir pressure, production pressure difference and liquid cushion density; relative density of gas-phase and crude oil, gas-oil ratio during production
Well structure	Well size, well depth, casing size, outer diameter and inner diameter of the casing, initial depth of casing, etc.
Well deviation	Sequence number of well sections, well deviation angle, azimuth angle, top depth, bottom depth, section length, etc.
Pipe column data	Sequence number of pipe column, pipe column size, outer diameter and inner diameter of pipe column, top depth, bottom depth, section length, remark, etc.
Temperature parameters	Seawater surface temperature, geothermal gradient, temperature calculation mode, measured temperature, etc.

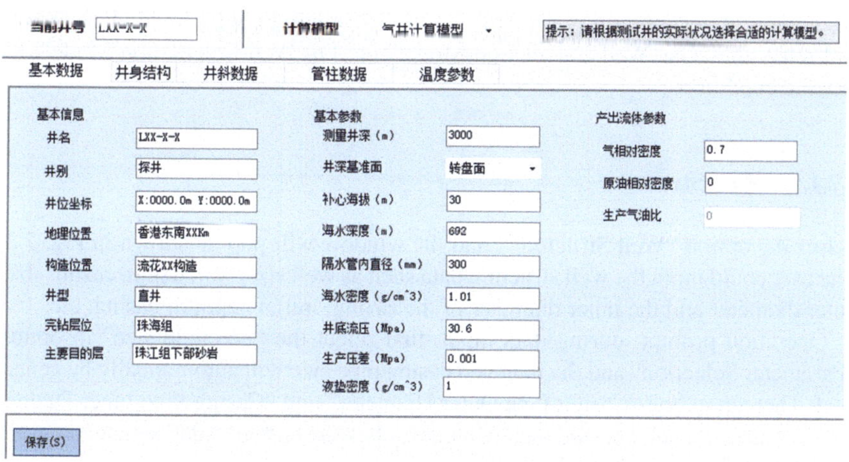

Fig. 7.2 Basic data on "Data Input" window

edit the basic information and parameters of the gas well. The basic information includes well name, well category, well position coordinates, geographic position, structure position, well type, drilled layers and main targeted layer. When moving the mouse to the input textbox, corresponding prompt information will appear. The basic parameters include well depth, well depth datum plane, kelly bushing, seawater depth, the inner diameter of water riser, seawater density, reservoir pressure, production pressure difference, and liquid cushion density. The user shall pay attention to the units of the corresponding parameter during date input. The parameters of produced

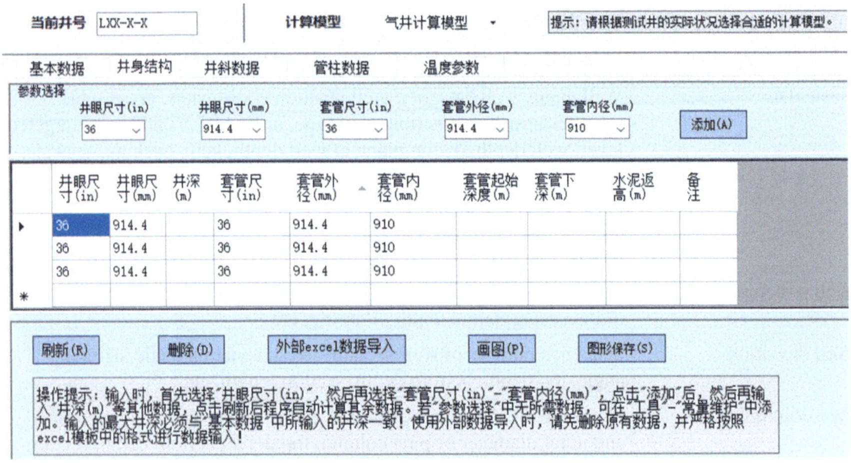

Fig. 7.3 Well structure data on "Data Input" window

fluid include the relative density of gas-phase and crude oil, gas-oil ratio during production. After inputting or modifying this information, we should click "Save" and select the "Gas Well Calculation Model" or "Oil Well Calculation Model" by specific conditions.

7.2.1.2 Well Structure

Select the option "Well Structure", and the window will pop up shown in Fig. 7.3. The user could input the well structure data such as well size, well depth, casing size, outer diameter and the inner diameter of the casing, initial depth of casing, etc.

Operation prompt: during data input, first select the "borehole size" in option "Parameter Selection" and the matched casing pipe size will automatically be generated. Thereupon, the "Casing Pipe Outer Diameter" and "Casing Pipe Inner Diameter" will automatically be updated. After that, click the button "Add" and the program will add a new row in the table. Besides, to ensure the input integrity, the user should input the lacked data on the new row. The user could also achieve the batch data processing by clicking the "Import the External Excel". Finally, the user could click "Draw" to visually display the well structure.

7.2.1.3 Well Deviation

Select the option "Well Deviation" and one parameter input and edit window shown in Fig. 7.4 will pop up. The user could input the well deviation data such as sequence number of well sections, well deviation angle, azimuth angle, top depth, bottom depth, section length.

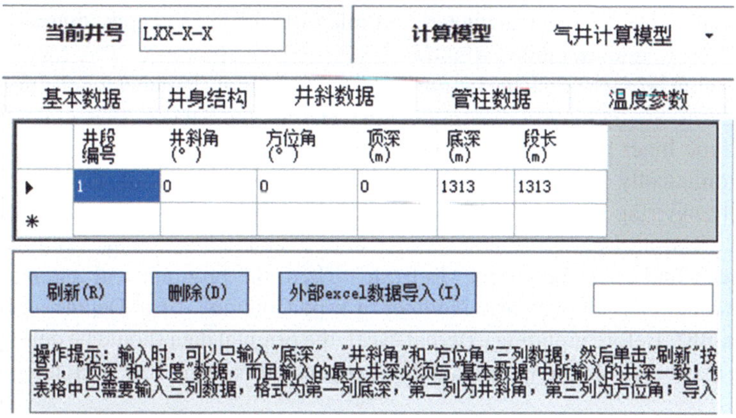

Fig. 7.4 Well deviation on "Data Input" window

Operation prompt: The user could input just three columns of data: "Bottom Depth", "Well Deviation Angle" and "Azimuth Angle", and click the button "Refresh" to automatically compute the data: "Sequence Number of Well Section", "Top Depth" and "Section Length". Before inputting external excel, the original data should be deleted first and the excel data should be collected and organized based on the form template.

7.2.1.4 Pipe Column Data

Select the option "Pipe Column Data" and the window shown in Fig. 7.5 will pop up. The user could input the pipe column data such as sequence number of pipe column,

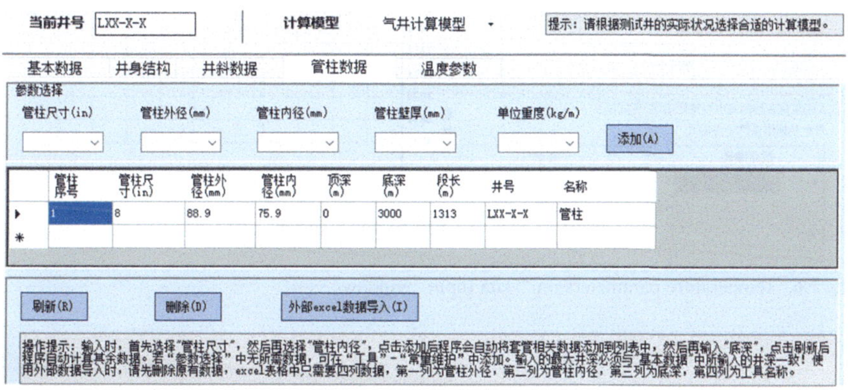

Fig. 7.5 "Pipe Column Data" on "Data Input" window

pipe column size, its outer diameter, and inner diameter, top depth, bottom depth, section length, remark on this window.

Operation prompt: Click the button "Add" in the option "Parameter Selection" and the program will add a new row in the table. Herein, the data of "Pipe Outer Diameter", "Pipe Inner Diameter" and "Pipe Column Size" are known, and the program will automatically add remaining data by clicking the button "Refresh". Also, when manually entering data in order, the user could only need to input three columns of data: "Bottom Depth", "Pipe Column Outer Diameter" and "Pipe Column Inner Diameter", and click the button "Refresh". Then, the program will automatically calculate the data of "Sequence Number of Pipe Column", "Top Depth" and "Section Length". Before inputting external excel, the original data should be deleted first and the excel data should be collected and organized based on the form template.

7.2.1.5 Temperature Parameters

Select the option "Temperature Parameters" and the "Temperature Parameters" window shown in Fig. 7.6 will pop up. Then, set the parameters such as temperature calculation mode, seawater surface temperature and geothermal gradient in the option "Temperature Parameters Input", and click the button "Display" after data saving. Herein, the ambient temperature will be displayed on the graph. If the users have related monitoring data, they could be updated in "Seawater and Formation Temperature Input".

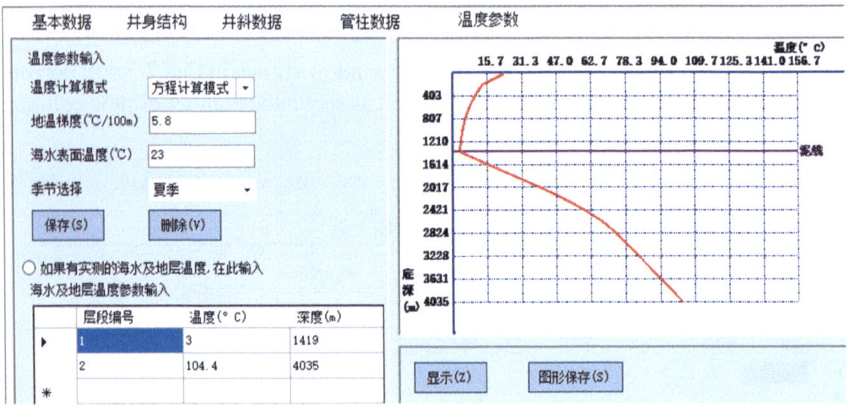

Fig. 7.6 Temperature parameters on "Data Input" window

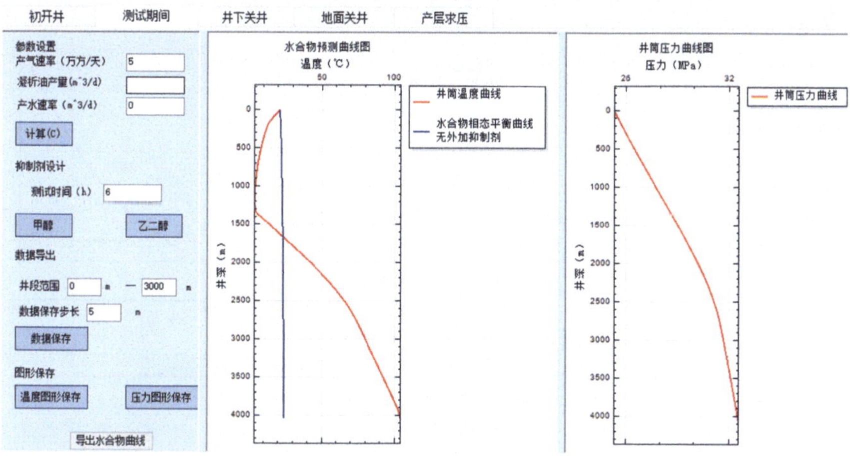

Fig. 7.7 Calculation module of well temperature and pressure fields

7.2.2 *Calculation Module of Well Temperature and Pressure Fields*

The calculation module of well temperature and pressure fields could be used to calculate the temperature and pressure distribution in gas well during well testing and production [5, 6], which is shown in Fig. 7.7. Using the well testing as an example, based on the gas well model, the temperature and pressure fields in pipe columns at different gas and water production rates can be calculated. Likewise, at different oil and water production rates and gas-oil ratios, the temperature and pressure fields in the oil well can be calculated.

7.2.3 *Calculation Module of Hydrate Phase Equilibrium Conditions*

The calculation module of the hydrate phase equilibrium conditions is shown in Fig. 7.8. In this module, it is necessary to first input the data including gas components, salt inhibitor, alcohol inhibitor, then set the auxiliary parameters including hydrate type and calculated temperature range [7].

It needs to note that before the gas components analysis, only methane is added to the gas component for the sake of safety. If the salt exists in output water or testing solution, corresponding salt will need to be added. After that, remember to click "Yes" in the option of "Combined Inhibitor Design" of "Auxiliary Parameters". In contrast, if "No" is selected, it means that in the process of judging of hydrate formation area, only gas is taken into account and the default values are taken in

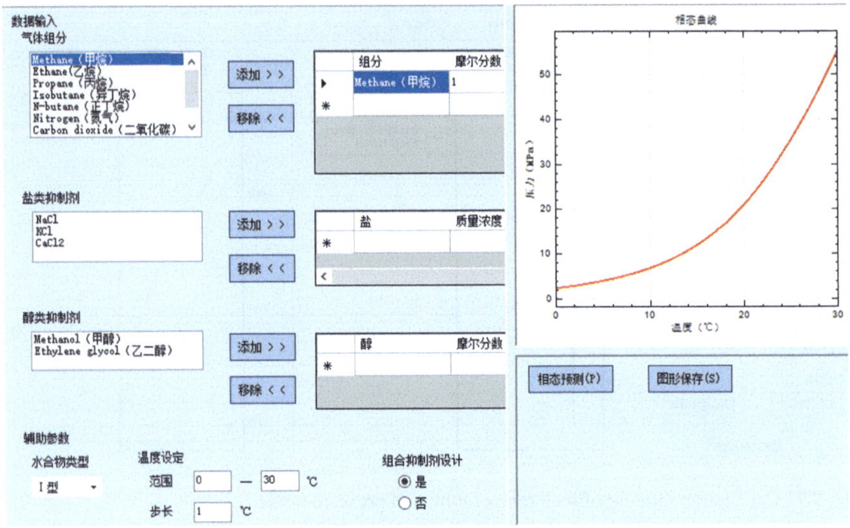

Fig. 7.8 Window of "Prediction for Hydrate Phase Equilibrium"

other parameters. Afterwards, clicking "Prediction for Hydrate Phase Equilibrium" and "Save the Image" could display the calculation results in a graph and save them separately. If the hydrate phase equilibrium is not carried out, only the temperature and pressure fields in the gas well will be obtained in the subsequent calculation. Herein, the hydrate formation area is not judged.

7.2.4 Prediction Module of Hydrate Formation Area

In this prediction module, after selecting the corresponding working conditions, the users would enter the flow of calculating the hydrate formation area and designing the methanol and ethanediol. Now, taking the well testing as an example, the corresponding operations are introduced and as shown in Fig. 7.9. The operation in other working conditions is similar to the well testing except for the input parameters.

During well testing, the user needs to set the gas and water production rates, then calculate and graphically display the well temperature and pressure curve and the hydrate phase equilibrium curve. These data could be saved as a text file and the related graphs could be saved separately. If hydrate formation occurs, by clicking the buttons "Methanol" and "Ethanediol" in the option of "Inhibitor Design", corresponding inhibitor design will be conducted and displayed. Otherwise, "No Hydrate Formation" will be prompted in "Inhibitor Design", so the buttons of "Methanol" and "Ethanediol" will be unavailable. What's more, these calculation results during well testing must be verified based on the measured wellhead temperature and pressure before the "engineering calculation".

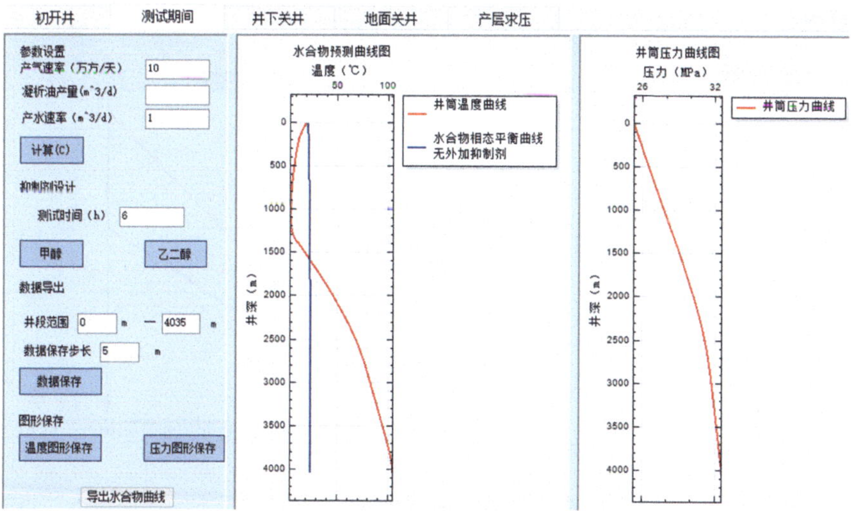

Fig. 7.9 "Hydrate Formation Area Prediction" window during test

Remarks: the "Computing " window would appear if the computation load is large during inhibitor design. Please wait patiently.

7.2.5 *Optimization Module of Injection Parameters*

The optimization module of injection parameters includes the calculation of injection parameters and chart for query during well testing. After obtaining the hydrate formation area, clicking "Inhibitor Injection Parameters" could get access to the "Single Inhibitor Injection Parameters" window shown in Fig. 7.10. The window includes three parts: injection parameters comparison, depth of injection point and injection rate during well testing.

Specifically, "injection parameters comparison" could display various parameters, such as working conditions, hydrate formation area, volume fraction of methanol and ethanediol, related remarks, in form of a chart. "depth of injection point" could be calculated according to corresponding working conditions. After setting the gas production, if hydrates are generated, the chart for alcohol inhibitor injection would be displayed. Otherwise, "No Hydrate Formation" will be prompted. Besides, by clicking the right key, the user may set the water content and injection rate, adjust the display area and format of the chart, which is shown in Fig. 7.11. Or the injection rate could be exported in form of table for further query.

Clicking the "Hydrate Management Chart in Deepwater Gas Well" could get access to the "Hydrate Management Chart in Deepwater Gas Well" window shown in Fig. 7.12. Firstly set "Output", "Alcohol Water Ratio" and "Oil Nozzle Diameter",

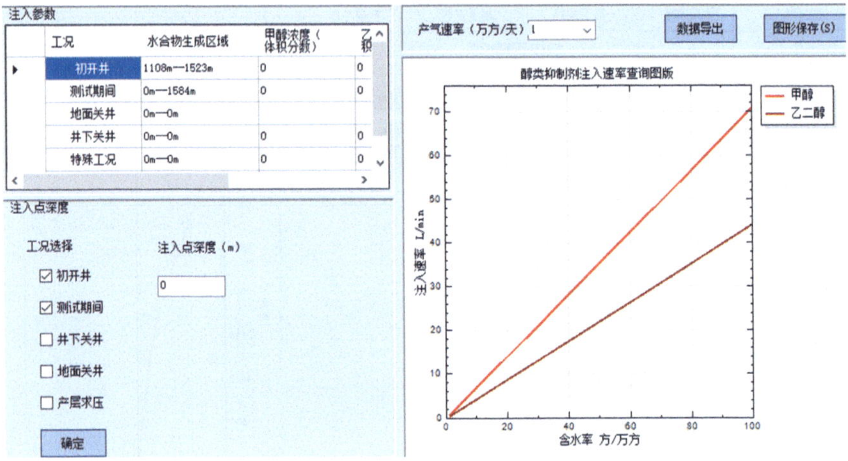

Fig. 7.10 “Single-inhibitor Injection Parameters” window

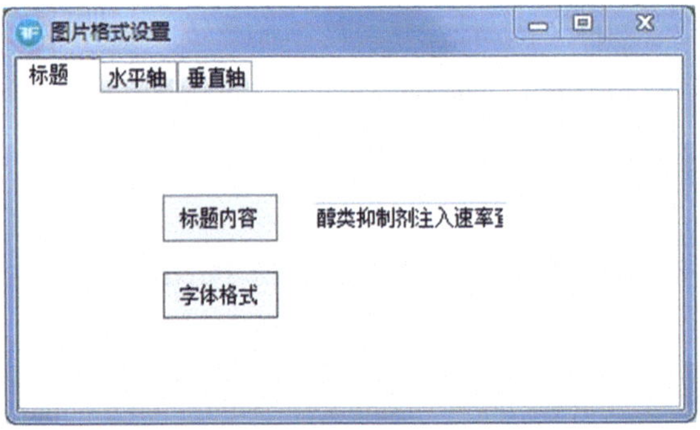

Fig. 7.11 “Image Format Setting” window

or recover the default parameters by clicking “Parameter Reset”, and then perform the corresponding calculation. After that, their results are displayed in graphs to facilitate the judgment of hydrate formation. Clicking “Methanol” or “Ethanediol” could display the corresponding hydrate management chart, which could also be saved for further query. For the other working conditions, their operation is similar to those of well testing.

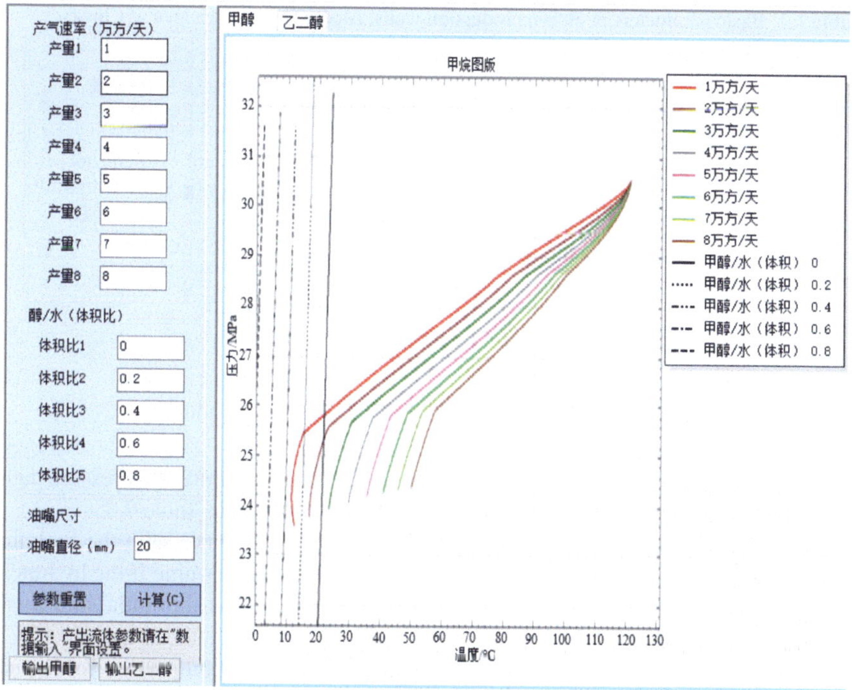

Fig. 7.12 "Hydrate Management Chart during Well Testing in Deepwater Gas Well" window

7.3 Case Analysis for Hydrate Management in Deepwater Gas Well

The influence of well temperature and pressure distribution, water depth, gas components, gas production and throttling effect on the hydrate formation area at different stages are analyzed according to the relevant data of the gas wells in block A of some certain sea area. Based on this, the severity and dimensions of the hydrate deposition and blockage are predicted and the inhibitory effect of inhibitors is analyzed comprehensively.

7.3.1 Basic Parameters

Based on the above description, the related hydrate prediction method is applied to eleven production wells in block A of one certain sea area to obtain the hydrate formation area during production, well shut-in and testing. The basic parameters of such production wells are shown in Table 7.2.

Table 7.2 Basic parameters of eleven production wells, Block A in one sea area of China

Natural gas components					
CO_2/%	C_1/%	C_2/%	C_3/%	*i*-C_4/%	*n*-C_4/%
3.4	87.8	5.80	2.1	0.4	0.5
Water depth (m)	Test position (m)	Well type	Pipe column outer diameter (mm)	Methanol density (g ml^{-1})	Methanol viscosity (mPa s) (20 °C)
1350	3150–3190	Straight well	114.3	0.79	0.59

7.3.2 *Prediction on Hydrate Formation Area*

(1) Prediction on Hydrate Formation Area during Normal Production

According to the designed production of gas wells, the temperature and pressure data in wellhead of different years could be calculated and shown in Table 7.3.

By comparing the well temperature and pressure fields curve with the hydrate phase equilibrium curve in the same proportion, we could determine if the hydrates are generated in the wellbore. Taking well #3 for example, at the initial and final periods of gas production, the temperature and pressure in wellhead are (60.1 °C, 33.2 MPa) and (29.3 °C, 5.8 MPa) respectively. Herein, the comparison results of the well temperature fields curve with the hydrate phase equilibrium curve in form of temperature-depth could be obtained and shown in Fig. 7.13. The results illustrate that at the initial and final periods of gas production, the temperature at each position of well 3# is higher than the hydrate formation temperature. So, there are no hydrate formation risks observed in well 3#. Likewise, by comparing the well temperature and pressure with the hydrate phase equilibrium condition during the normal production period of remaining years, it has been found that there are no hydrate formation risks in well #3 all along.

For the other wells, the wellhead temperature at the initial and final periods of gas production is shown in Table 7.4, illustrating the wellhead temperature at the initial and final periods of gas production are in the range of 70.5–81.3 °C and 24.6–42.3 °C respectively. Similarly, the well temperature field curves are disjoint from the hydrate phase equilibrium curve in form of temperature-depth for each well, indicating that there are still no hydrate formation risks in all the wellbore at the initial and final periods of production.

(2) Prediction on Hydrate Formation Area under Well Shut-in Condition

Both the gas production and bottom hole flowing pressure change over time, so the well temperature and pressure are distributed variously with the well shut-in times. Taking well #2 for example, the well temperature and pressure distribution at the initial period of well shut-in is shown in Fig. 7.14, illustrating that the wellbore temperature gradually decreases with time and finally approaches to the stratum temperature, but its decreasing speed gradually slows down with time. This

Table 7.3 Temperature and pressure data in wellhead

Time	Well #1		Well #2		Wel #3		Well #4		Well #5		Well #6	
	Temperature/°C	Pressure/ MPa	Temperature/°C	Pressure/ MPa	Temperature/°C	Pressure/ MPa	Temperature/°C	Pressure/ MPa	Temperature/°C	Pressure/ MPa	Temperature/°C	Pressure/ MPa
2020	76.1	28.6	80.6	29.3	60.1	33.2	73.8	30.6	75.1	30.3	76.6	28.7
2021	76.3	25.4	80.9	24.6	60.0	29.9	73.8	28.9	75.1	28.8	76.8	24.3
2022	76.6	22.2	81.2	21.2	59.8	27.3	73.9	27.2	75.2	27.2	77.1	21.2
2023	76.8	19.8	79.2	20.3	62.7	24.8	77.5	23.4	78.7	23.3	74.4	20.2
2024	77.3	17.6	79.4	18.7	62.7	22.6	77.9	21.3	79.1	21.3	74.5	18.6
2025	77.8	15.6	79.7	16.8	62.6	20.6	78.6	18.7	79.5	19.4	74.8	16.6
2026	78.7	13.5	80.2	14.9	62.7	18.8	79.7	15.6	79.8	18.1	75.3	14.5
2027	80.0	11.3	80.9	12.9	62.8	16.9	81.4	12.7	81.1	15.2	76.0	12.5
2028	81.6	9.4	82.2	10.8	63.2	13.2	82.2	11.6	82.7	12.3	77.4	10.2
2029	81.9	7.3	82.6	8.3	63.2	13.1	82.8	9.9	82.9	11.2	80.5	7.5
2030	80.2	5.5	83.2	6.6	63.6	11.3	83.5	8.5	83.0	9.5	81.9	6.0
2031	74.8	7.9	78.9	8.0	64.3	9.2	79.5	7.9	83.5	5.8	77.2	5.7
2032	73.5	8.3	78.2	7.7	66.1	6.8	82.6	7.5	81.1	5.5	76.7	5.4
2033	72.8	7.5	79.9	6.3	74.2	3.3	75.9	7.4	79.0	6.6	75.8	4.6
2034	70.7	7.1	80.6	5.1	63.8	4.0	69.2	7.5	73.6	8.0	75.1	4.0

(continued)

Table 7.3 (continued)

Time	Well #1		Well #2		Well #3		Well #4		Well #5		Well #6	
	Temperature/°C	Pressure/MPa	Temperature/°C	Pressure/MPa	Temperature/°C	Pressure/MPa	Temperature/°C	Pressure/MPa	Temperature/°C	Pressure/MPa	Temperature/°C	Pressure/MPa
2035	69.1	6.8	80.3	4.4	63.3	3.6	69.2	7.5	73.4	5.9	65.7	4.7
2036	69.5	6.0	78.2	4.2	59.9	3.2	67.6	7.1	73.0	5.3	59.3	5.1
2037	70.9	4.9	74.1	4.5	53.5	3.7	66.6	6.5	71.4	4.9	54.1	5.4
2038	76.8	3.1	73.2	4.2	47.2	4.3	65.4	6.0	70.8	4.3	50.1	5.5
2039	63.8	4.3	69.9	4.3	43.1	4.6	64.2	5.4	70.4	3.7	45.5	5.6
2040	64.1	4.2	66.6	4.4	38.7	5.2	62.9	4.8	68.7	3.5	45.6	5.3
2041	59.8	4.2	62.9	4.7	36.6	5.4	63.6	3.7	60.5	5.5	41.5	5.2
2042	59.3	3.9	59.6	4.8	31.8	5.7	53.7	5.8	57.9	6.0	36.8	5.3
2043	60.0	3.2	56.4	4.9	29.3	5.8	47.0	6.9	62.8	3.7	31.4	5.5
2044	56.8	3.1	50.2	5.2			38.7	7.6	51.0	6.2		
2045	52.7	3.4	47.4	5.3					49.7	6.0		
2046	48.9	3.6	44.1	5.4					47.4	6.7		
2047	46.2	3.7	39.2	5.5					45.4	6.8		
2048	43.4	3.7	36.3	5.5					43.4	6.9		
2049	41.9	3.8										

(continued)

Table 7.3 (continued)

Time	Well #7		Well #8		Well #9		Well #10		Well #11	
	Temperature/°C	Pressure/MPa	Temperature/°C	Pressure/MPa	Temperature/°C	Pressure/MPa	Temperature/°C	Pressure/MPa	Temperature/°C	Pressure/MPa
2020	81.6	27.9	78.1	31.6	72.3	29.0	71.0	29.5	74.3	29.8
2021	82.0	22.9	78.1	30.2	72.3	28.1	71.0	28.6	74.4	27.2
2022	82.5	19.3	78.1	28.7	72.3	26.9	71.1	27.3	74.6	25.0
2023	79.6	18.4	80.2	26.2	75.5	23.8	75.0	23.8	77.8	21.2
2024	79.9	16.7	80.3	24.4	74.6	21.9	75.2	21.8	78.3	18.9
2025	80.4	14.6	80.5	22.7	76.0	20.1	75.5	20.0	79.0	16.8
2026	81.3	12.5	80.7	20.7	76.4	18.3	75.9	17.9	79.3	15.5
2027	82.9	10.2	80.9	18.9	77.0	16.3	76.6	15.6	80.9	12.7
2028	83.3	7.7	81.4	17.0	78.6	12.9	77.9	12.9	81.4	11.3
2029	83.9	6.6	82.2	14.5	81.8	9.1	79.8	10.4	81.3	9.8
2030	82.5	6.3	83.2	11.5	82.2	6.7	80.4	7.8	82.7	7.3
2031	81.9	6.0	83.5	10.2	83.1	4.0	75.1	8.5	79.7	6.3
2032	82.4	5.5	83.8	8.7	80.5	4.4	73.1	9.3	78.0	6.2
2033	82.6	5.1	82.3	7.0	77.8	5.3	69.3	9.3	73.6	6.8
2034	82.5	3.5	77.4	7.4	72.5	5.8	70.0	7.9	70.7	6.8

(continued)

Table 7.3 (continued)

Time	Well #7		Well #8		Well #9		Well #10		Well #11	
	Temperature/°C	Pressure/MPa	Temperature/°C	Pressure/MPa	Temperature/°C	Pressure/MPa	Temperature/°C	Pressure/MPa	Temperature/°C	Pressure/MPa
2035	73.0	4.6	73.0	7.5	71.5	5.4	67.6	7.2	66.1	7.4
2036	66.6	5.3	69.3	7.5	79.2	2.3	65.1	7.1	64.6	7.1
2037	61.7	5.6	66.8	7.1	71.0	3.1	64.2	6.3	64.3	6.5
2038	57.3	5.9	64.5	6.7	63.7	3.6	63.9	5.1	65.6	5.5
2039	52.1	6.0	61.9	6.4	58.3	4.1	67.2	3.3	66.0	4.7
2040	48.6	6.1	59.0	6.0	55.3	4.1	62.7	3.4	66.6	4.0
2041	43.1	6.2	56.7	5.3	51.9	4.3	60.2	3.3	73.6	2.5
2042	39.9	6.2	55.0	4.6	48.3	4.5	56.5	3.4	67.7	2.6
2043	36.4	6.3	48.4	4.9	44.6	4.8	52.2	3.9	59.7	3.5
2044	32.5	6.3	45.2	5.1	40.5	5.2	46.4	4.4	53.8	4.3
2045	25.8	6.4	41.7	5.1	38.5	5.5	46.4	4.4	48.3	4.7
2046			37.9	5.1			44.4	4.6	44.6	5.0
2047			35.7	5.0			40.5	4.9	40.7	5.3
2048			33.4	5.0			40.4	4.8	31.7	5.7
2049			31.0	5.0			38.3	4.9		

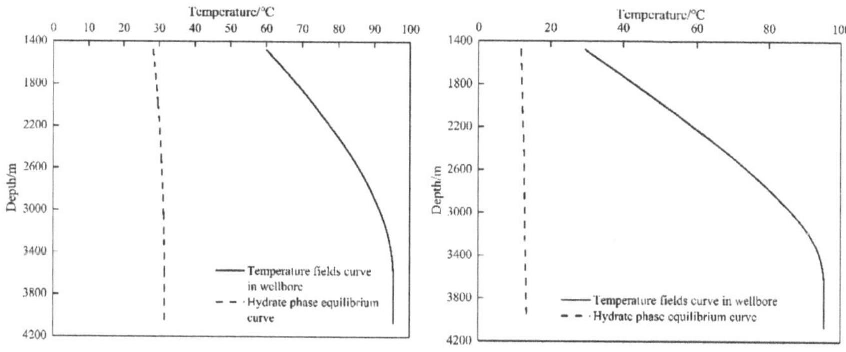

(a) Initial production period (b) Final production period

Fig. 7.13 Prediction on hydrate formation area in Well #3

decreasing process is lasting for about 29.7 h due to the decreasing rate of heat exchange between wellbore fluid and stratum. It is also found that the wellbore pressure gradually reduces from the bottom to the top of wellbore caused by the own gravity of the gas.

Before the well shut-in, these wells have different ages of gas production. Herein, the time when the wellbore temperature falls to the ambient temperature after shut-in is shown in Fig. 7.15, illustrating that the longer gas production time, the shorter time when the wellbore temperature falls to the ambient temperature after shut-in is. This is because the longer production time induces, the lower wellbore temperature and pressure is, and results in the less heat contained in the gas. Consequently, it is easier for time to decrease the ambient temperature.

After well shut-in, by comparing the well temperature fields curve with the hydrate phase equilibrium curve, the hydrate formation area in well #2 is shown in Fig. 7.16. It illustrates that the well temperature starts falling to the hydrate formation temperature after shut-in of 4.7 h, and the hydrate formation area gradually becomes larger with the decrease of wellbore temperature. And, the hydrate formation area in well #2 would reach a maximum when the wellbore temperature is equal to the stratum temperature. Herein, the shut-in time reaches 29.7 h and the area is ranging from mudline to the 457 m below mudline. It is also found that the smaller distance off the mudline induces, the larger radial distance of the hydrate formation area is. That is to say, the larger subcooling degree could lead to higher risks of hydrate formation. On the contrary, the larger distance off the mudline means the lower hydrate risks. Finally, the hydrate risks would disappear at 457 m below mudline.

Under the different production years before well shut-in, the time when the hydrate formation area appears in the wellbore after shut-in is shown in Fig. 7.17. It shows that the longer gas production time is, the shorter time is when the hydrate formation area appears in the wellbore after shut-in is. This is because the longer production time brings the lower wellbore temperature and pressure, so that the time when the wellbore temperature falls to the ambient temperature after shut-in is shorter,

Table 7.4 Wellhead temperature of each well at the initial and final period of gas production

	Well #1	Well #2	Well #3	Well #4	Well #5	Well #6	Well #7	Well #8	Well #9	Well #10	Well #11
Temperature at the initial period /°C	75.8	80.3	60.1	73.3	74.6	76.3	81.3	77.7	71.8	70.5	73.9
Temperature at the final period/°C	41	35.3	29.3	37.5	42.3	30.2	24.6	29.9	37.5	37.2	31.6

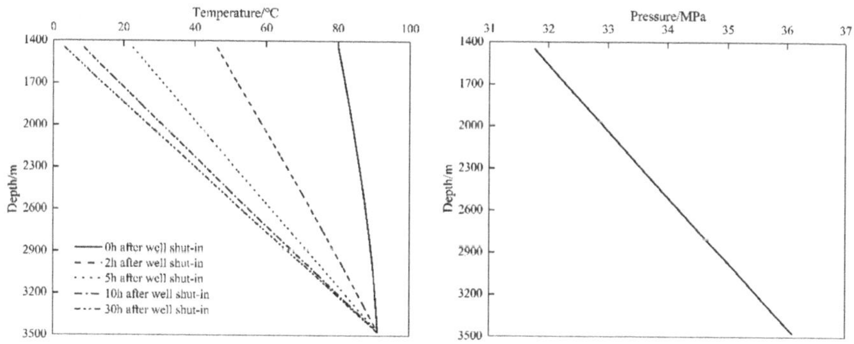

(a) Temperature Distribution

(b) Pressure Distribution

Fig. 7.14 Changes of wellbore temperature and pressure during shut-in of Well #2

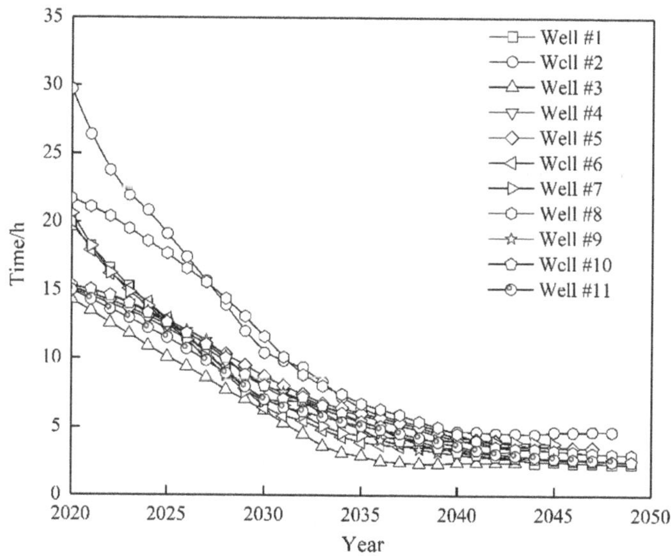

Fig. 7.15 Time required for the well temperature falling to ambient temperature during well shut-in with different production years

which is easier for well fluid to meet the conditions for hydrate formation. The detailed data of the hydrate formation area is shown in Table 7.5, illustrating that there are different hydrate formation areas in the wellbore after wells shut-in. Therefore, specific measures for hydrate management need to be taken according to the specific characteristics of each well.

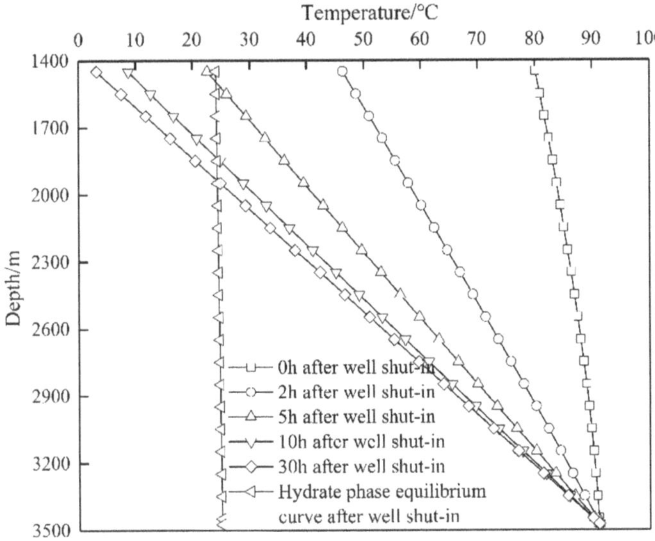

Fig. 7.16 Prediction on hydrate formation area after well #2 shut-in

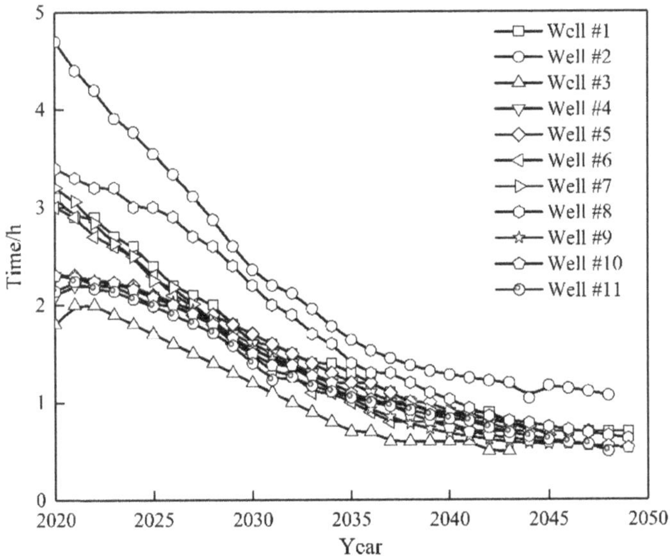

Fig. 7.17 Time when the hydrate formation area appears after well shut-in

Table 7.5 Hydrate formation area of each well in Block A after shut-in

	Well #1	Well #2	Well #3	Well #4	Well #5	Well #6	Well #7	Well #8	Well #9	Well #10	Well #11
Hydrate formation area/m	1336–1819	1477–1934	1447–1903	1547–1978	1547–1975	1531–1991	1466–1919	1466–1936	1252–1743	1252–1747	1365–1837

(3) Prediction on Hydrate Formation Area during Well Testing

During well testing, the prediction results on hydrate formation area in eleven wells with different gas productions are shown in Fig. 7.18, illustrating that under almost all the gas production, there exists the hydrate formation risk in the eleven wells.

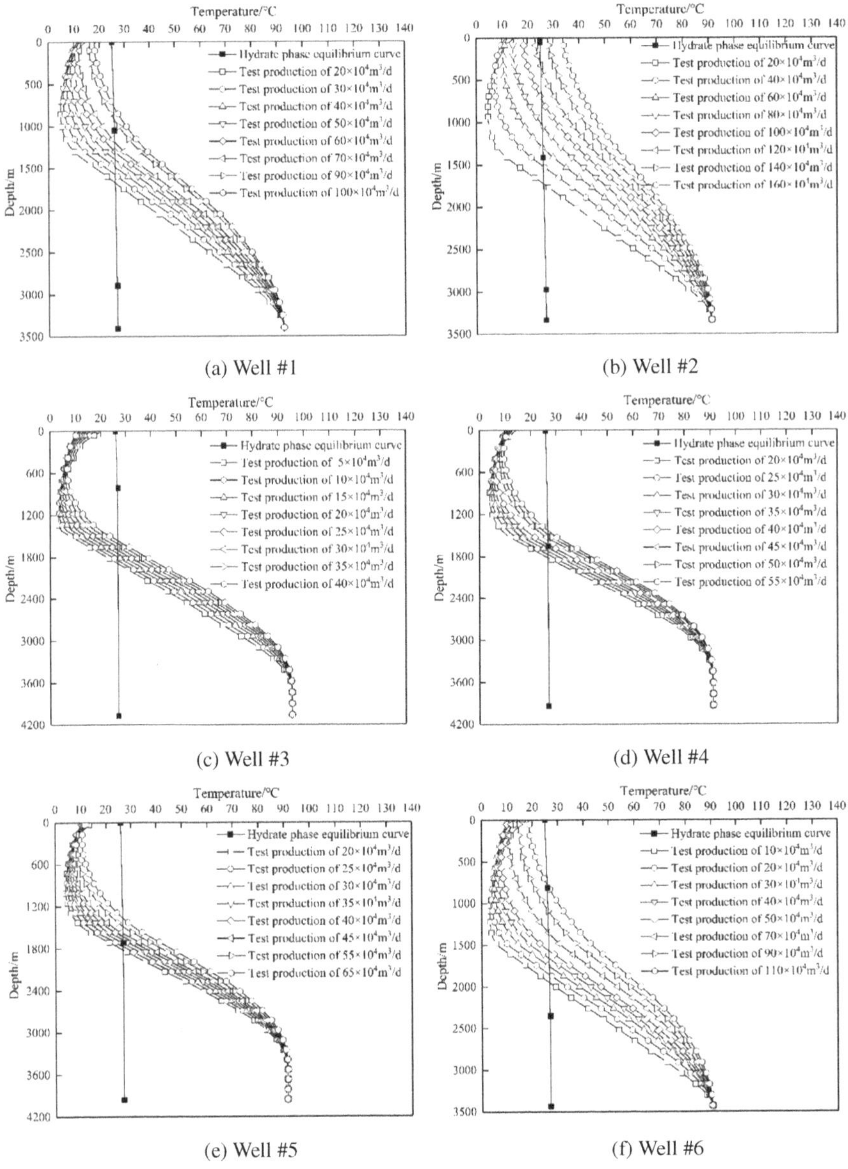

Fig. 7.18 Prediction on hydrate formation area in each well during well testing

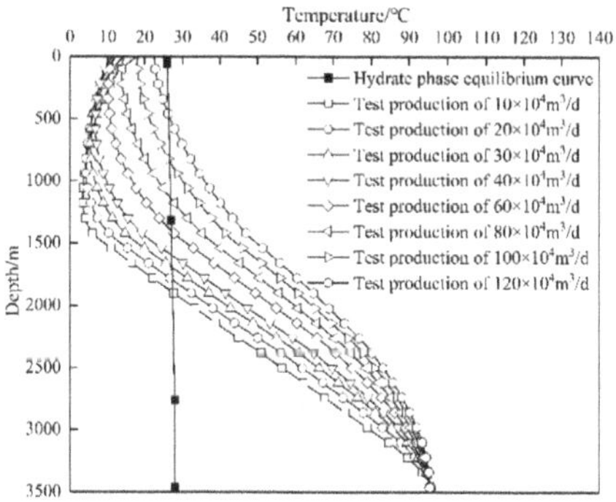

(g) Well#7

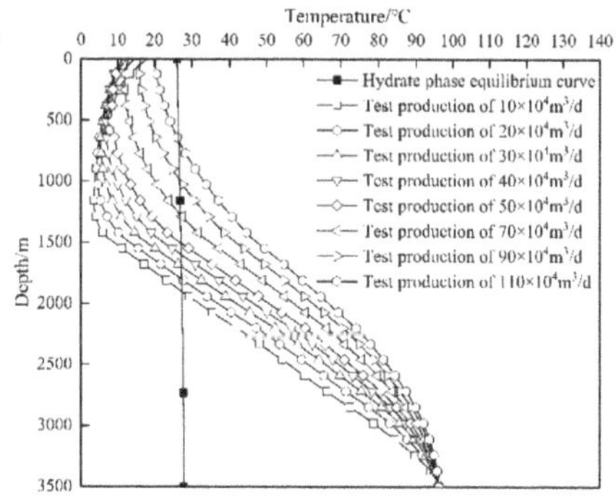

(h) Well #8

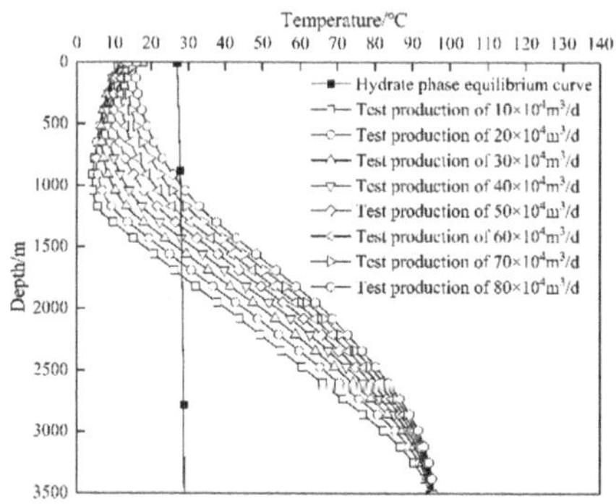

(i) Well #9

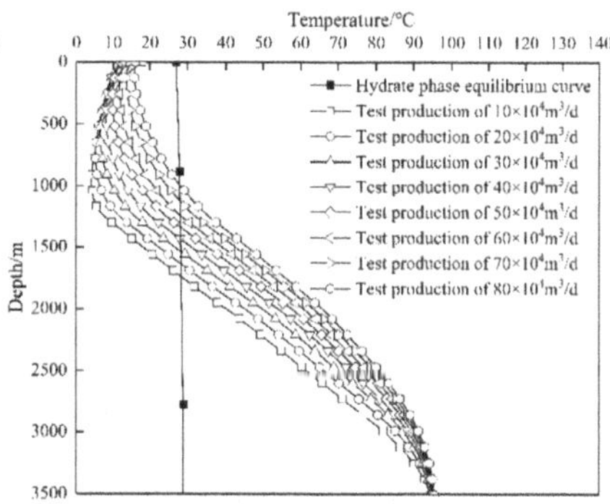

(j) Well #10

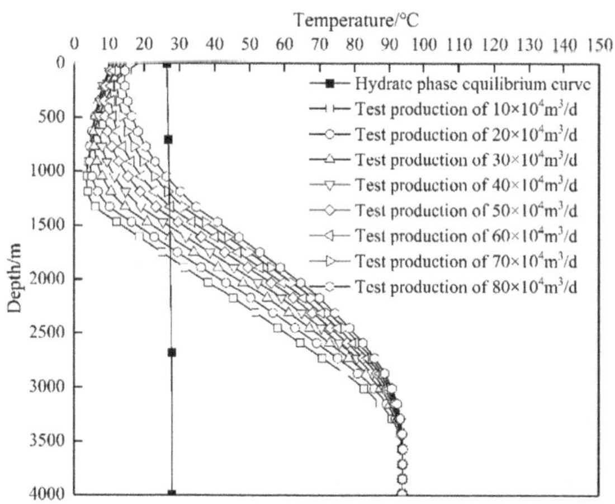

(k) Well#11

Fig. 7.18 (continued)

Besides, the hydrate formation area is mainly distributed in the seawater section. In the seawater section with low temperature, the heat exchange rate between well fluid and ambient temperature is fast. Thus, the wellbore in the seawater section is easy to reach the hydrate phase equilibrium conditions. Furthermore, the hydrate formation area in well becomes smaller with the gas production during well testing.

7.3.3 *Prediction on Hydrate Blockage*

(1) Prediction on Hydrate Blockage after Shut-in

Well #2 in block A is taken for example to describe the evolution of hydrate deposition and blockage after well shut-in. The hydrate layer growth in well #2 after shut-in at downhole and surface shut-in is shown in Fig. 7.19, illustrating that whatever the way of a shut-in is, the longer shut-in time would result in the larger value of δ_D due to the continuous accumulation of the hydrate layer. Specifically, δ_D reaches a maximum after shut-in at downhole for 30 h. Moreover, the δ_D distribution in well is not uniform at the same time [3]. The larger value of δ_D could be observed near the wellhead due to the large subcooling degree at the position.

(2) Prediction on Hydrate Blockage during Well Test

The hydrate layer growth in well #2 during testing is shown in Fig. 7.20, illustrating that δ_D gradually increases over the test time. Furthermore, the lower gas production, the larger the fluctuation hydrate layer grows. From the shrink diameter ration of the wellbore, we could find there exists one extreme point of δ_D. This is because the distribution of hydrate formation area changes at different productions, and the process of hydrate formation and deposition are also influenced by subcooling degree and system temperature [7].

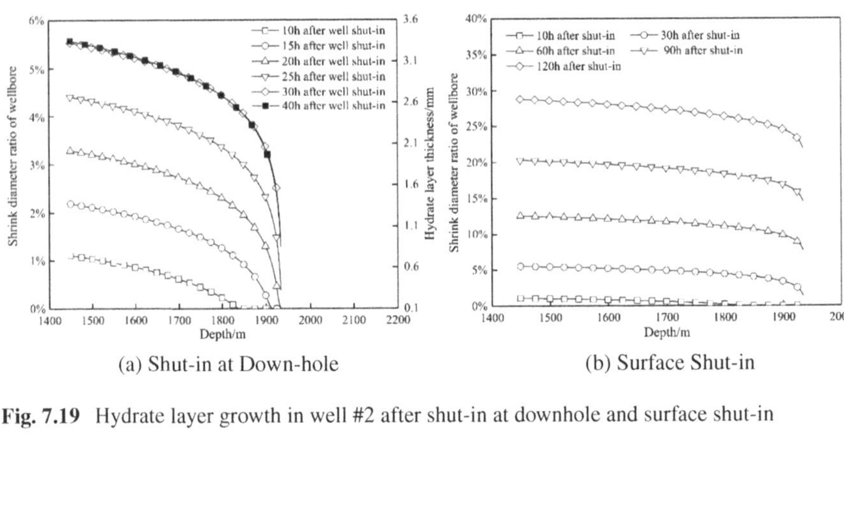

(a) Shut-in at Down-hole (b) Surface Shut-in

Fig. 7.19 Hydrate layer growth in well #2 after shut-in at downhole and surface shut-in

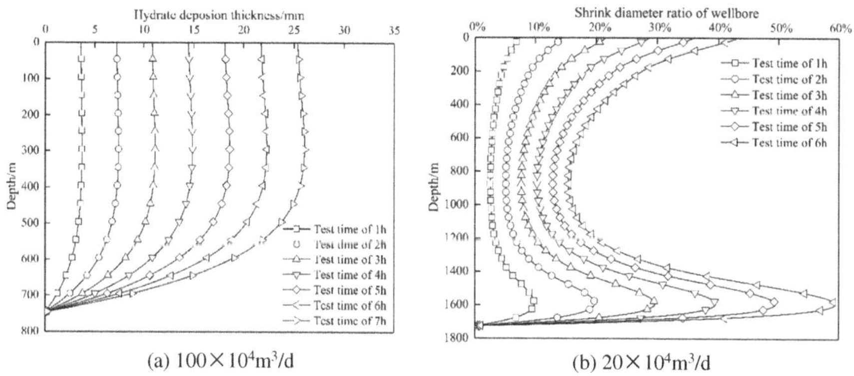

Fig. 7.20 Hydrate layer growth during testing of well #2

In well #2, the changes of shrink diameter ratio under the gas production of 120×10^4 m^3/d and the changes of wellhead backpressure under different gas productions are shown in Fig. 7.21 and Fig. 7.22 respectively, illustrating the shrink diameter ratio of wellbore increases with the testing time, and its maximum in well #2 is only 11% when well testing is lasting for 2 h. Also, the larger gas production, the faster wellhead backpressure drops. Hence, the changes of wellhead backpressure could help field staff judge if the hydrates are generated in the wellbore and guide the injection design of inhibitor.

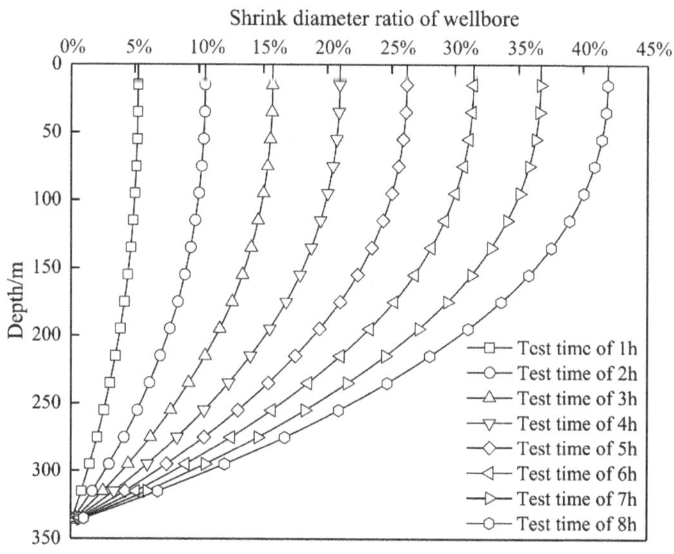

Fig. 7.21 Shrink diameter ratio of well #2 under the gas production of 120×10^4 m^3/d

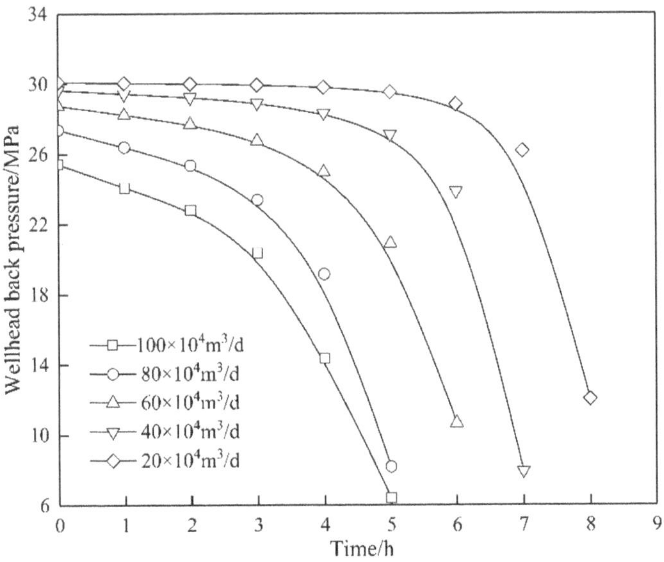

Fig. 7.22 Change of wellhead back pressure in well #2 under different productions

7.3.4 Chart for Injection Concentration

According to the basic parameters of eleven production wells, the well temperature and pressure distribution at different gas productions and hydrate phase equilibrium curve at different concentrations of alcohol inhibitors are calculated respectively. Based on the intersection area of two curves, we could visually judge if the hydrates are formed in one certain production and inhibitor concentration, finally determining the injection concentration of inhibitors under different gas productions.

Now, well #1 is taken as an example to describe the chart of injection concentration during the injection of methanol and ethanediol. This chart is shown in Fig. 7.23. It illustrates that the well temperature and pressure curve intersects the hydrate phase equilibrium curve in different productions, and hydrate formation would occur without any inhibitor injection. As the increase of gas production and injection concentration, both two curves are away from each other, and the hydrate formation area in wellbore becomes smaller. If no intersection points are found, no hydrate will be generated in the wellbore. Based on this, the injection concentration of methanol and ethanediol in certain production could be judged to prevent and manage the hydrate blockage.

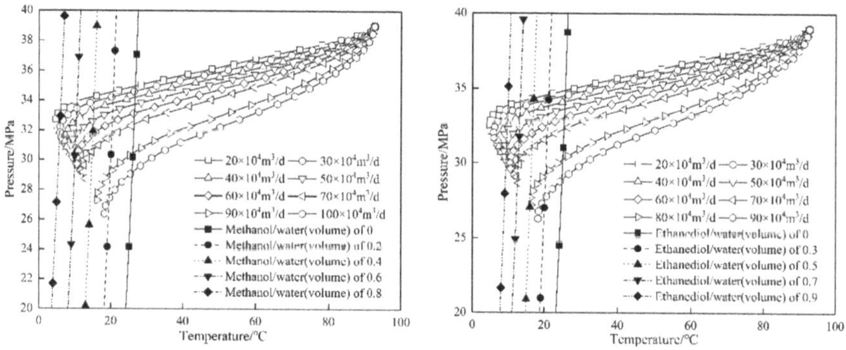

(a) Methanol (b) Ethanediol

Fig. 7.23 Chart for injection concentration during injecting the methanol and ethanediol into well #1

7.3.5 *Chart for Injection Pressure*

The injection pressure is closely related to the gas and water production during testing and shall be accurately calculated according to the specific conditions. Now, an analysis of the injection pressure of well #8 at different gas and water productions is conducted.

Under different gas productions and water contents, the chart for injection pressure during injection the methanol and ethanediol into well #8 is shown in Fig. 7.24, illustrating that the inhibitor injection pressure increases first and then increase. This is because with the increase of gas production, the inhibitor injection concentration decreases and water content increases. The common influence of such two factors determines the developing trend of injection pressure. Furthermore, in the same gas

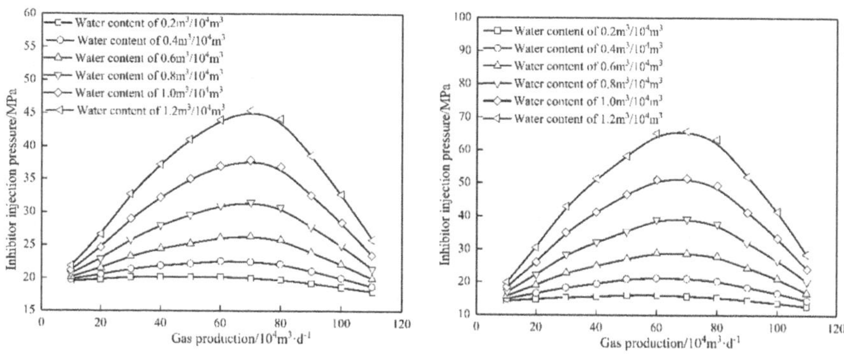

(a) Methanol (b) Ethanediol

Fig. 7.24 Chart for injection pressure during injecting the methanol and ethanediol into well #8

production, both the injection amount and injection pressure gradually increase with the water content.

It is also found that because the ethanediol with higher density and viscosity would cause higher friction pressure drop at high water content, the injection pressure of ethanediol is higher than that of methanol. By contrast, at low water content, the injection pressure of ethanediol is lower than that of methanol because the injection amount of ethanediol is small when the water content is a little low, resulting in the little friction pressure drop. Hence, ethanediol is to be preferred at low water content or methanol at high water content during well testing.

References

1. China University of Petroleum (East China). Design software for hydrate prevention during deepwater gas well testing: China. 2014SR145750.2014.07.01[CP/OL]
2. Wang Z, Yang Z, Zhang J et al (2018) Quantitatively assessing hydrate-blockage development during deepwater-gas-well testing. SPE J 23(4):1166–1183
3. Wang Z, Zhao Y, Zhang J et al (2018) Flow assurance during deepwater gas well testing: hydrate blockage prediction and prevention. J Petrol Sci Eng 163:211–216
4. Wang Z, Zhao Y, Sun B et al (2016) Modeling of hydrate blockage in gas-dominated systems. Energ Fuel 30(6):4653–4666
5. Wang Z, Zhao Y, Sun B et al (2018) Features and prevention of gas hydrate blockage in test strings of deep-water gas wells. Nat Gas Ind 38(01):81–88
6. Wang Z, Zhang J, Jiang H et al (2017) Modeling and application of wellbore multiphase flow with gas hydrate phase transition. Chin J Hydrodyn 32(5): 584–591
7. Wang Z, Sun B, Wang X et al (2014) Prediction of natural gas hydrate formation region in wellbore during deep-water gas well testing. J Hydrodyn Ser. B 26(4):568–576

The manufacturer's authorised representative in the EU is Springer Nature Customer Service Centre GmbH, Europaplatz 3, 69115 Heidelberg, Germany. If you have any concerns regarding our products, please contact ProductSafety@springernature.com

Printed and bound by CPI Group (UK) Ltd, Croydon, CR0 4YY
30/09/2026
02225557-0001